W9-DJJ-893

CONTEMPORARY ISSUES IN TOURISM DEVELOPMENT

This work combines a study of contemporary issues in tourism development with close examination of approaches to tourism research. Looking beyond much-studied mass tourism industries, leading international academics explore new issues raised by emerging tourist destinations such as Ghana, Samoa, Vietnam and India's Bhyundar Valley.

Subjects covered include:

- reasons for development
- tourism development as a strategy for urban revitalisation
- tourism's links to heritage conservation and regional development
- sustainability and the adverse impacts of development
- cultural considerations and community participation
- the importance of context for individual tourism projects

In methodological chapters, contributors challenge the current thinking in tourism studies. A tension is identified between calls for more generalized, theoretical approaches to research and the more case-oriented approach stressing contextual factors. This book proposes alternative approaches to research and policy issues.

Tourism Development is at the cutting edge of a rapidly developing field and will be a valuable resource for all those interested in the changing face of tourism and tourism studies.

Douglas G. Pearce is Associate Professor of Geography at the University of Canterbury, Christchurch, New Zealand. **Richard W. Butler** is Professor in the School of Management Studies for the Service Sector at the University of Surrey, UK. They have co-edited two previous works for Routledge: *Tourism Research: Critiques and Challenges* (1993) and *Change in Tourism: People, Places, Processes* (1995).

ROUTLEDGE ADVANCES IN TOURISM
Series Editors: Brian Goodall and Gregory Ashworth

1 THE SOCIOLOGY OF TOURISM
Theoretical and Empirical Investigations
Edited by Yiorgos Apostolopoulos, Stella Leivadi and Andrew Yiannakis

2 CREATING ISLAND RESORTS
Brian King

3 DESTINATIONS
Cultural Landscapes of Tourism
Greg Ringer

4 MEDITERRANEAN TOURISM
Facets of Socioeconomic Development and Cultural Change
Edited by Yiorgos Apostolopoulos, Lila Leontidou, Philippos Loukissas

5 OUTDOOR RECREATION MANAGEMENT
John Pigram and John Jenkins

6 CONTEMPORARY ISSUES IN
TOURISM DEVELOPMENT
Edited by Douglas G. Pearce and Richard W. Butler

CONTEMPORARY ISSUES IN TOURISM DEVELOPMENT

Edited by Douglas G. Pearce and Richard W. Butler

London and New York
in association with the
International Academy for the Study of Tourism

First published 1999
by Routledge
11 New Fetter Lane, London EC4P 4EE

Simultaneously published in the USA and Canada
by Routledge
29 West 35th Street, New York, NY 10001

Routledge is an imprint of the Taylor & Francis Group

Typeset in Garamond by
Exe Valley Dataset Ltd, Exeter
Printed and bound in Great Britain by
Mackays of Chatham plc, Chatham, Kent

British Library Cataloguing in Publication Data
A catalogue record for this book is available from the British Library

Library of Congress Cataloging-in-Publication Data
Tourism development: contemporary issues/edited by Douglas G. Pearce and Richard W. Butler.
p. cm.
Includes bibliographical references and index.
1. Tourist trade. 2. Tourist trade—Research—Methodology.
I. Pearce, Douglas G., 1949– . II. Butler, Richard, 1943– .
G155.A1T59114 1999
338.4'791'072—dc21 98-52802
CIP

ISBN 0-415-20691-X (hbk.)

CONTENTS

FIGURES

TABLES

CONTRIBUTORS

Richard W. Butler
School of Management Studies for the Service Sector, University of Surrey, Guildford, United Kingdom

Graham M.S. Dann
Department of Travel, Tourism and Leisure, University of Luton, Luton, United Kingdom

William R. Eadington
Institute for the Study of Gambling and Commercial Gaming, University of Nevada, Reno, Nevada, USA

William C. Gartner
Tourism Center, University of Minnesota, St Paul, Minnesota, USA

Sang Mu Kim
Department of Tourism Management, Keimyung University, Taegu, Korea

Myriam Jansen-Verbeke
Department of Social and Economic Geography, Catholic University Leuven, Belgium

Carson L. Jenkins
The Scottish Hotel School, University of Strathclyde, Glasgow, Scotland

Els Lievois
Department of Social and Economic Geography, Catholic University Leuven, Belgium

Gianna Moscardo
School of Business, James Cook University, Townsville, Australia

Douglas G. Pearce
Department of Geography, University of Canterbury, Christchurch, New Zealand

Philip L. Pearce
School of Business, James Cook University, Townsville, Australia

John J. Pigram
Centre for Water Policy Research, University of New England, Armidale, NSW, Australia

Linda K. Richter
Department of Political Science, Kansas State University, Manhattan, Kansas, USA

Barbara A. Rugendyke
School of Geography, Planning, Archaeology and Palaeoanthropology, University of New England, Armidale, NSW, Australia

Regina Schlüter
Centro de Investigaciones y Estudios Turísticos, Buenos Aires, Argentina

Patricia Simpson
Department of Geography, University of Waterloo, Waterloo, Ontario, Canada

Shalini Singh
Graduate School of Business and Administration, Ghaziabad, India

Tej Vir Singh
Centre for Tourism Research and Development, Lucknow, India

Nguyen Thi Son
School of Geography, Planning, Archaeology and Palaeoanthropology, University of New England, Armidale, NSW, Australia

Geoffrey Wall
Faculty of Environmental Studies, University of Waterloo, Waterloo, Ontario, Canada

1

INTRODUCTION

Issues and approaches

Douglas G. Pearce

INTRODUCTION

Throughout the world, tourism continues to develop as a significant social and economic activity. World Tourism Organization figures indicate that receipts from international tourism more than doubled in the decade 1988–97, increasing from US$204 billion to US$444 billion (WTO 1998). International arrivals grew at a slower rate but in absolute terms rose from 394 million in 1988 to 613 million in 1997. Less comprehensive figures are available for domestic tourism. Figures for individual Western countries show experience in the late 1980s and early 1990s varied considerably. In some, such as France, domestic holidays continued to increase; elsewhere (for example, the United Kingdom, Germany, Australia and New Zealand), signs of stagnation or decline were appearing, in part as a consequence of increased outbound tourism (Pearce 1995). In contrast, domestic tourism was becoming more significant in some developing countries as changing standards of living meant domestic travel was becoming more accessible to a wider cross-section of society.

Much tourism remains geographically concentrated: just over half of the international arrivals (52 per cent) and the receipts (54 per cent) in 1997 were recorded in the ten leading destinations. Countries in Europe and North America are still major sources and destinations for international tourism but in recent years have been joined by others, especially in Asia and the Pacific. Varying rates of growth have been recorded in these and in other regions of the world, such as Africa and South America. Resorts in some of the latter regions have a relatively long tradition of tourism, such as Bariloche in Argentina. In other places, for example parts of Africa or some Pacific islands, tourism is still emerging as a form of development. While numbers there may as yet be relatively insignificant on a global scale, they may be of increasing significance locally. As a consequence of these patterns and processes, tourism is either already widely established as a major sector of national, regional and local economies in many parts of the world or being actively considered or fostered as a development option in many others.

The growth of tourism has been accompanied by a significant increase in research and scholarly activities in this field. Tourism programmes, centres and departments have multiplied on campuses worldwide, tourism conferences have proliferated and new publications on tourism mushroomed. Symptomatic of this trend is the proliferation of new and increasingly specialized tourism journals; those that have appeared in the 1990s include: *Journal of Tourism Studies* (1990), *Estudios y Perspectivas en Turismo* (1991), *Journal of Travel and Tourism Marketing* (1991), *Journal of Sustainable Tourism* (1993), *Tourism Economics* (1995), *Progress in Tourism and Hospitality Research* (1995), *Journal of International Hospitality, Leisure and Tourism Management* (1997), *Pacific Tourism Review* (1997), *Tourism Analysis* (1997).

Growth in the amount of research on tourism has not necessarily been matched by a concomitant increase in the quality of research being done in this field. While progress has been made over the last decade, many of the criticisms raised by Dann, Nash and Pearce in 1988 in terms of the degree of methodological sophistication and the lack of a strong theoretical base might still be applied today (Mowforth and Munt 1998). Not only is there a growing need to study tourism further as new issues and problems appear and the increasing magnitude of the sector enlarges its significance – socially, economically, environmentally and politically – but there is also a need to understand the phenomenon better by addressing the quality of the research that is being done and how that might be improved.

These interrelated issues constituted the focus of the fifth biennial meeting of the International Academy for the Study of Tourism held in Melaka, Malaysia, in June 1997 in association with the Ministry of Culture, Arts and Tourism, the Universiti Kebangsaan Malaysia and the State Government of Melaka. The theme of the Melaka meeting, from which this book results, was 'Tourism development: issues for a new agenda'. The Academy was established in 1988 as a limited-membership, multidisciplinary international body to 'further the scholarly research and professional investigation of tourism, to encourage the application of the findings, and to advance the international diffusion and exchange of tourism knowledge'. In its focus on contemporary issues in tourism development and ways of addressing these, the Melaka conference built on the four previous meetings of the Academy which had examined:

- alternative forms of tourism (Smith and Eadington 1992);
- methodological and conceptual issues in tourism research (Pearce and Butler 1993);
- change in tourism (Butler and Pearce 1995);

- environmental and community issues in tourism development (Cooper and Wanhill 1997).

In particular, this volume combines a focus on contemporary issues in tourism development with consideration of underlying methodological and theoretical matters.

A volume such as this cannot hope to be comprehensive in its coverage and exhaustive in its treatment; nor does it pretend to be. However, the international and multidisciplinary background of the Academy and of the contributors to this volume enables a range of issues to be addressed from a variety of disciplinary perspectives (geography, sociology, economics, political science, tourism studies) and illustrated with research in a range of geographical settings from Las Vegas and Leuven in the developed world to Vietnam and Ghana in the developing. It is hoped readers will appreciate both the diversity and eclecticism of the collection, which draws attention to a variety of new development and research issues, and also the underlying direction and coherence sought by the conference organizers and editors as a means of enabling broader themes and concerns to emerge from the selected revised papers.

The focus here is not on more familiar, well-established, mass tourism destinations such as the coastal resorts of the Mediterranean or the alpine resorts of Europe (see Montanari and Williams (1995) for a recent review of these). Rather, a feature of this volume is the inclusion of material from parts of the world where tourism is just developing and little research has previously been carried out (e.g. Ghana, Samoa, Vietnam, and India's Bhyundar Valley) and other regions such as Patagonia which have a longer tradition of tourism but where hitherto there has been little material available in English. Gartner (Chapter 10), for example, demonstrates quite clearly in his analysis of Small Scale Enterprises (SSEs) in Ghana's Central Region how new issues may arise, both in terms of development and the research which supports it, when attempts are made to develop tourism in a new destination. He finds that classical theories do not deal adequately with SSEs and that to understand local development issues cultural conditions must be injected into the equation. Conversely, Schlüter (Chapter 11) argues that much of what is portrayed in the literature as new in terms of ecotourism has its parallels in much more established forms of tourism in Patagonia. Other contributors adopt a much more explicit methodological or philosophical stance by directly considering the ways in which tourism research is done, how it might be done better and how new, more theoretically based, approaches might be pursued. Dann (Chapter 2), for example, advocates the development of a stronger theoretical base for the study of tourism and outlines ways this might be accomplished while Pearce and Moscardo (Chapter 3), in their analysis of work on tourism community relations, emphasize the need to sharpen up the research focus and to ask the right questions. Jenkins (Chapter 4) directly addresses a range of methodological issues associated with the implementation of tourism development plans and Jansen-Verbeke and Lievois (Chapter 6) outline some of the methodological issues associated with the development of heritage tourism.

To place the individual chapters in context and make more explicit the broader themes which they contain, the next two sections outline two interrelated sets of concerns: contemporary tourism development issues and approaches to research in this field.

ISSUES

A key issue in any discussion of tourism development, one which underlies the patterns outlined in the increase in the WTO figures cited above and one which runs in a variety of ways through the chapters which follow, is why is tourism growing, and why is its development being encouraged? This issue can essentially be addressed from either a demand or a supply-side perspective. Demand or origin studies stress changes in market conditions which affect people's motivation to travel and the factors which influence their ability to do so, for example increased leisure time and disposable income, improved technology and travel organization (Pearce 1995). Supply-side or destination research, the focus of the contributions in this book, tends to address the benefits that the development of tourism brings or is perceived to bring, to consider what leads both the public and private sectors to foster its growth and how this might best be achieved. The benefits of tourism development have largely been seen in economic terms, as in tourism's ability to generate income, jobs and corporate profits, bring in foreign exchange, boost tax revenues, diversify the economy and aid regional development. Social and environmental benefits have also been recognized (Pearce 1989). As more traditional sectors of the economy, first agriculture and then manufacturing, have come under pressure regarding their ability to deliver in these terms, so tourism has been increasingly targeted as a potential alternative or complement in an increasingly wide range of settings from the urban to the rural, from developed to developing countries. The different ways in which this occurs is well illustrated by the contributions in this volume.

Increasingly, the development of tourism is being linked to other processes. Jansen-Verbeke and Lievois (Chapter 6) show how tourism development has been identified within the European Union as a strategy for urban revitalization, a strategy which is tied in to other policies promoting heritage conservation in the historic cities of Europe. However, as the authors note, such supply-side policies are only made feasible by the growing interest in such features as a consequence of a broader cultural revival. A similar growth in heritage tourism and in the number of registered historical sites in the

United States is recorded by Richter (Chapter 7). This rapidly increasing demand is attributed to a variety of factors including alienation and nostalgia while the development of ever more sites is seen to be a community response to tourism pressure, that is controlling growth in one place by opening alternative sites elsewhere. In a completely different context, Eadington (Chapter 8) outlines how tourism development has been used as one of the key rationales behind the spread of casino gambling over the past decade. Casino gambling, he argues,

> has been legalized primarily as a means to achieve other 'higher purposes' such as tourism development, partly because the activity itself is considered – especially by policy makers – to be of questionable merit. Justifications are not found in the demand for gambling from the general public, but rather in the economic spin-offs that are thought to occur when gambling is authorized.

Whether or not such spin-offs actually materialize, especially when casinos are opened in urban areas, is more debatable, but the extent to which the tourism growth argument has held perhaps illustrates just how far tourism has come in being recognized as a development strategy. What is clear is that where these links are being made between tourism development and other processes, be they heritage and nature conservation, the legalization of gambling or regional development, sound research is needed to explore the claims being made and where appropriate to contribute to the implementation of the policies in question.

A common theme in the developing country chapters is that while the question of sustainability is recognized and the need to mitigate any adverse social, economic and environmental impacts is acknowledged, 'the emphasis is laid on development' (Schlüter, Chapter 11) and 'that development must and will occur' (Simpson and Wall, Chapter 14). Issues here thus relate to ways of achieving development while mitigating potential adverse impacts. Son, Pigram and Rugendyke (Chapter 13), for example, discuss the use of national parks as a way of developing nature-based tourism, often advanced as a 'less predatory and more benign form of tourism', but note that nature conservation and the encouragement of tourism within national parks do not necessarily or readily go hand in hand and particular attention is needed to balance one function against the other. Schlüter also points out in the case of Patagonia that political and territorial motives may underlie the creation of national parks and the fostering of tourism in frontier regions. Other contributors stress the need to take into account cultural considerations, whether in terms of the organization of public sector involvement in the case of Samoa (Pearce, Chapter 9), encouraging private sector initiatives in Ghana (Gartner, Chapter 10) or community participation in the Bhyundar Valley in the Himalayas (Singh and Singh, Chapter 12).

In stressing that cultural and environmental considerations should be taken into account, the developing country chapters also draw attention to the need to consider processes of tourism development and the way in which different forms of tourism generate different levels and types of impact (Pearce 1989, 1992a). This point is addressed in general terms by Simpson and Wall in Chapter 14 and elaborated on by other contributors in the context of their study areas. Schlüter discusses how many development models in South America have frequently been imported and imposed and makes a case for more localized responses. Increasing local participation is generally seen as being desirable and ways of analysing and fostering this are outlined in various developing world settings in the chapters dealing with Samoa, Ghana, Vietnam and the Bhyundar Valley .

As noted above, knowledge of the potential or actual impacts generated is a key factor in decisions as to what forms of development are appropriate and how much tourism is desirable. Pearce and Moscardo demonstrate in Chapter 3 that although the assessment of tourism impacts is a large and well-established mainstream component of the literature there is still considerable scope to refine assessment techniques, and they argue that this might best be achieved by developing a sounder theoretical approach to the problem. In their complementary examination of actual practice in Indonesia, Simpson and Wall highlight not only the need for appropriate Environmental Impact Assessment (EIA) procedures for tourism development projects to be formulated but also that these should be subsequently implemented in a timely and effective manner if they are to have any influence on subsequent outcomes. Their case study in North Sulawesi reveals a considerable gap between theory and practice. Kim (Chapter 15) tackles a novel issue by assessing the impacts of Korean tourists on Korean nationals engaged in tourism operations in two international destinations, Hawaii and Queensland, thereby adding a new dimension to the analysis of the globalization of tourism development.

Finally, many of the contributors point to the need to take more account of the context of tourism development. While common features may be identified in terms of processes and impacts, crucial differences are also to be found from one place to another. Successful tourism development, it is argued, will depend on a full appreciation of contextual factors and the way in which these are incorporated into the development process. This general point is well made by Butler (Chapter 5) in his systematic overview of the problems and issues of integrating tourism development. Butler adopts a broad meaning of integration, namely:

> We can regard integrated planning and development as meaning the process of introducing tourism into an area in a manner in which it mixes with existing elements. It is implicit in such an understand-

> ing that this introduction and mixing is done in an appropriate and harmonious way, such that the end result is an acceptable and functionally successful community, in both ecological and human terms.

Desirable as such a development process may be, it is not one which is without its difficulties, as he then goes on to discuss.

Eadington too underlines the importance of contextual factors, observing that while Las Vegas is frequently portrayed as the exemplar of tourism and gambling in the United States, because of its unique circumstances it 'has set a somewhat unattainable ideal' for other jurisdictions. He also shows how the ratio of benefits to costs steadily declines as one moves across a continuum for casinos or casino-style gaming from resorts to neighbourhood locations or even the home.

Pearce and Singh and Singh caution against the use of a single or narrow range of development models. Pearce reiterates an earlier finding (Pearce 1992b: 200) that 'There is no single best type of [tourist] organization nor interorganizational network, rather each country must evolve a system which best reflects local, regional and national conditions' before examining how an appropriate national tourist organization for Samoa might be structured which takes into account the *faaSamoa*, the traditional Samoan way of life. Likewise, Singh and Singh note limitations on the suitability of transferring the Sherpa model of tourism development to the Bhotias of the Bhyundar Valley, noting 'no single indigenous model can be a perfect fit as each unique indigenous culture is constantly evolving in the face of change in the environment in which it exists'.

APPROACHES

The question of the general versus the specific which comes through in the review of substantive development issues also pervades much of the discussion on approaches to research on tourism development. The early chapters in the book address these approaches to research explicitly and in more general terms while in the latter case-oriented chapters methodological matters arise more indirectly but are no less significant for this. If a certain tension and apparent inconsistencies appear between the different approaches advocated or adopted, this is perhaps a reasonable reflection of the state of the art. However, the juxtaposition of varying approaches does highlight the need to be more explicit about the way we are approaching our research and the strengths and weaknesses of each, and provides the opportunity to examine these questions more systematically.

Dann (Chapter 2) offers a provocative critique of tourism research claiming:

> Instead of there being a desirable cumulative corpus of knowledge that is emic, comparative, contextual and processual (Cohen 1979), what we frequently encounter is a ragged collection of half-baked ideas that constitutes largely descriptive, case-confined wishful thinking. Moreover, in the few instances where there is an optimal balance between theoretical awareness and methodological sophistication (Dann *et al.* 1988), the sheer diversity of disciplinary approaches may well mean that their various representatives are rarely speaking the same paradigmatic language.

The way forward, he argues, lies in going beyond description to place greater emphasis on theory and on developing understanding. Three broad approaches by which this might be pursued are outlined: the Toffler or Futures approach, the Simmelian perspective and open-ended work. Particular emphasis is given to the latter, where a variety of strategies are suggested under the headings: blind alleys, self-appropriation, reversing conventional wisdom, concept stretching, scope broadening, breaking out of the case, resolving paradoxes and establishing new linkages. Each of these strategies provide opportunities for generating new research questions and stimulating our thinking about tourism development.

Pearce and Moscardo (Chapter 3) are also concerned with the issue of how researchers generate their studies. While their focus is on tourism community relations, the matters they discuss are relevant to other areas of tourism development and indeed to tourism research in general. After noting that the setting of tourism research agenda is a frequently repeated activity in which issues such as sustainability and quality often arise, Pearce and Moscardo observe: 'One of the challenges for researchers examining these research agenda is how to proceed from these generic directions to the development of insightful and incisive research.' Their answer lies in the questions researchers ask. Thus they consider how we should ask the right questions. Issues they identify here relate to the level and depth of the question, the need for a clear focus and the challenge of relevance. After a substantial literature review of tourism community relationships they conclude that:

> We need to reformulate this area of study; we need to ask new questions or develop a new theoretical perspective so that we do not pursue an endless litany of unconnected studies using different definitions which fail to provide a cumulative body of knowledge.

The theoretical approach they employ is social representations research.

Similar concerns are expressed by Simpson and Wall who preface their discussion of EIA for tourism by stating:

> In recent decades there has been a proliferation of studies of the impacts of tourism. Many of the findings of these studies have been contradictory and the case study approaches which have often been adopted have yet to lead to the cumulative knowledge or level of generalization desired by decision makers.

At the same time, Simpson and Wall, like many of the other contributors mentioned earlier, acknowledge the need to take account of contextual factors which may create significant differences from one setting or case to another.

Faced with calls for building a cumulative body of knowledge about tourism development while also taking account of contextual factors, the challenge confronting tourism researchers today is to adopt approaches which respect the latter but also contribute to the former. The emphasis given to each will depend in large part on the specific purpose of any particular study. Where the primary goal is to develop theory then the range of strategies proposed by Dann and Pearce and Moscardo become the central focus. The need to 'break out of the case', for example, becomes paramount. In terms of policy-supporting research, the level of generalization required by decision makers needs to be carefully considered. It is unrealistic (and indeed may be unnecessary) to envisage that all decisions will be based on case-specific research, but the limitations of applying policies and models from outside as discussed by Eadington, Schlüter and other contributors must be recognized. In other instances, the emphasis in policy-supporting research will first be on developing appropriate techniques and methodologies. The specific patterns found in the heritage resources of Leuven may not be repeated in other historic cities but the techniques outlined in Chapter 6 by Jansen-Verbeke and Lievois to identify them will have application elsewhere. And of course theories, as well as more general ideas, normative or otherwise, need to be tested against empirical evidence as Richter observes in Chapter 7.

The chapters in this book illustrate several ways forward in terms of resolving the tension between calls for more generalized, theoretical approaches to research on tourism development and those which have adopted a more case-oriented approach stressing contextual factors. The most common is to ensure that the analysis of any particular case is set squarely in the literature, the real-world relevance of the problem is explicitly elaborated and the more generalized applications and implications of the specific findings are outlined. This facilitates aspects of the cumulative knowledge-building process discussed by Dann. Good examples here include the chapters dealing with Samoa, Ghana, Vietnam and Indonesia.

In the case of Samoa, explicit provision for contextual or environmental factors is built into the open-systems inter-organizational model employed. While this facilitates evaluation of the national tourist organization in the context of the *faaSamoa,* the conceptual framework used and the linkage back to comparative studies on tourist organizations elsewhere also enables more general questions to be identified from this study of an emerging destination. Thus the study concludes:

> Further comparative work is now needed in order to ascertain the degree to which the features described are specific to Samoa or common to other small developing countries. How, for example, are cultural considerations expressed in the actions of other NTOs? Are there other ways in which NTOs can foster local participation in tourist development? To what extent do the forms of external aid determine the way in which NTOs in small developing countries are structured and function or in what ways is overseas assistance conditioned by the NTO's policies and activities? Are institutional vacuums common? How are they filled elsewhere?

The more general case for greater use of the comparative approach for advancing our understanding of tourism has been made by this author elsewhere (Pearce 1993).

Another useful strategy is to recognize the validity and complementarity of different approaches. Richter's review of the politics of heritage tourism is a good example of 'concept broadening'. She observes:

> Ordinarily social scientists do not think of tourism as political socialization and communication because it takes place largely outside the familiar institutions of politics: the home, church, school, media and government. Heritage destinations may convey particular political messages – intended and unintended – that have been only rarely studied.

Richter then goes on to review a range of issues, noting throughout the research gaps which appear, before concluding: 'Case studies are needed to probe who gains and loses from heritage tourism. There are a lot of assumptions about it with very little empirical data.'

A challenge of a different sort is offered by Jenkins (Chapter 4) in his stimulating review of the methodology of tourism development planning. Jenkins identifies a 'great divide' between academics and tourism practitioners, lays stress on the question of implementation rather than theory-building and suggests that much of the research undertaken by academics is less effective than it might otherwise be as they often give insufficient attention to the practical application of their ideas and findings. He then goes on to discuss the different stages in the tourism planning process and to consider ways in which the gap may be bridged and a better 'cross-over' achieved. Likewise Butler discusses both principles and issues of practicality with regard to integrating tourism development. Other chapters provide evidence that the 'cross-over' sought by Jenkins is possible and already occurring: those on Samoa and Ghana, for example, are written by academics involved in practical aid projects while that by Jansen-Verbeke and Lievois stresses methodological aspects of policy-supporting research.

CONCLUSIONS

The seemingly relentless growth of tourism will continue to create new issues in tourism development and reinforce the importance of others that have been apparent for some time. As a consequence, related research needs will also continue to grow. These needs are both quantitative and qualitative. Calls for increased theoretical understanding must be matched by further consideration of methodological matters and the need for practical implementation. The chapters in this volume do not cover all possible issues and approaches but individually and collectively they will contribute to extending our knowledge of tourism development, both through the examples they provide and by stimulating others to respond to the challenges outlined.

Acknowledgements

The editors wish to acknowledge the contribution of Kadir Din who acted as local organizer for the Melaka meeting of the Academy, and the invaluable assistance of Linda Harrison, Tim Nolan and James Guard of the Department of Geography, University of Canterbury, in preparing the final manuscript, in drawing or redrawing many of the figures in this volume and in converting the files and e-mails which came in many different versions into a usable format.

References

Butler, R. and Pearce, D.G. (eds) (1995) *Change in Tourism: People, Places, Processes*, London: Routledge.

Cohen, E. (1979) 'Rethinking the sociology of tourism', *Annals of Tourism Research* 6 (1): 18–35.

Cooper, C. and Wanhill, S. (eds) (1997) *Tourism Development: Environmental and Community Issues*, Chichester: Wiley.

Dann, G., Nash, D., and Pearce, P. (1988) 'Methodology in tourism research', *Annals of Tourism Research* 28 (2): 1–28.

Montanari, A. and Williams, A.M. (1995) *European Tourism: Regions, Spaces and Restructuring*, Chichester: Wiley.

Mowforth, M. and Munt, I. (1998) *Tourism and Sustainability: New Tourism in the Third World*, London: Routledge.

Pearce, D.G. (1989) *Tourist Development*, 2nd edn, Harlow: Longman, and New York: Wiley.
—— (1992a) 'Alternative tourism: concepts, classifications and questions', pp. 15–30 in V.L. Smith and W.R. Eadington (eds) (1992) *Tourism Alternatives: Potentials and Problems in the Development of Tourism*, Philadelphia: University of Pennsylvania Press.
—— (1992b) *Tourist Organizations*, Harlow: Longman and New York: Wiley.
—— (1995) *Tourism Today: A Geographical Analysis*, 2nd edn, Longman: Harlow.
—— (1993) 'Comparative studies in tourism research', pp. 20–36 in D.G. Pearce and R.W. Butler (eds) (1993) *Tourism Research; Critiques and Challenges*, London: Routledge.
Pearce, D.G. and Butler, R.W. (eds) (1993) *Tourism Research; Critiques and Challenges*, London: Routledge.
Smith, V.L. and Eadington, W.R. (eds) (1992) *Tourism Alternatives: Potentials and Problems in the Development of Tourism*, Philadelphia: University of Pennsylvania Press.
WTO (1998) *Tourism Highlights 1997*, Madrid: World Tourism Organization.

2

THEORETICAL ISSUES FOR TOURISM'S FUTURE DEVELOPMENT

Identifying the agenda

Graham M.S. Dann

INTRODUCTION

Tourism development, as Pearce (1989) rightly observes, is an expression that encompasses not only destinations, origins, motivations and impacts, but also the complex linkages that exist between all the people and institutions of that interlocking, global supply and demand system. He also notes that tourism development is a hybrid term, that research in the field consists of two essentially separate literatures and that there are few examples of good research that manage to combine the two domains successfully. One reason why scholars may be disenchanted with the merger is that both tourism and development lack individual ideological neutrality, indeed they can on occasion become highly emotive political issues (see Richter Chapter 7) and when considered jointly, either as state or process, they may become equally or even more value laden (Wall 1997). Such a quality is unlikely to attract academics trained in the rigours of impartiality and objectivity. Another, and arguably more significant, reason is that tourism development, like the domain of tourism itself, may still be at an early stage of theoretical advancement. Again, as Pearce (1989) illustrates, much of the intellectual treatment of tourism development has apparently not progressed much further than the stages of classification and model building of two decades ago, a situation enabling him to declare 'the field of tourist development is yet to be fully supported by a strong theoretical base' (Pearce 1989: 23). Echoing this sentiment, Butler (1997: 121) more recently claims that:

> The process of development of tourist destinations has received little attention in the literature, and conceptualization of the process has been very limited. There have been a large number of case studies of

> the pattern of development of destinations, but they have been based on a shallow theoretical foundation.

Considered in these terms, it is unlikely that a field in its theoretical infancy will be able to act as an oracle for the provision of reliable and timely advice to those governments, practitioners and communities depending on tourism for their livelihood (see Jenkins Chapter 4).

We may therefore have reached a stage where, if policy makers are often accused of being insufficiently proactive towards the ever-expanding phenomenon of tourism, the same charge can also be legitimately levelled against theoreticians. Most of the time, it would seem, the hallmarks of theory – understanding, explanation, prediction and falsifiability (Dann *et al.* 1988) – are at best piecemeal features of tourism research (Pearce 1982), only catching up with reality, if at all, once trends and events have already taken place. Instead of there being a desirable cumulative corpus of knowledge that is emic, comparative, contextual and processual (Cohen 1979a), what we frequently encounter is a ragged collection of half-baked ideas that constitutes largely descriptive, case-confined wishful thinking. Moreover, in the few instances where there is an optimal balance between theoretical awareness and methodological sophistication (Dann *et al.* 1988), the sheer diversity of disciplinary approaches may well mean that their various representatives are rarely speaking the same paradigmatic language.

However, while agreeing with many of the foregoing state-of-the-art observations, it is argued here that the above scenario may be a trifle pessimistic, since there are at least three adoptable, future-oriented positions towards tourism development which permit viable theoretical anticipation and reasonable progress. They may be designated 'The Toffler or Futures Approach', 'The Simmelian Perspective' and 'Open-Ended Work'. After briefly outlining the first two, this chapter concentrates on the third approach, exploring a variety of open-ended research strategies.

THE TOFFLER OR FUTURES APPROACH

Those familiar with the major works of Toffler – *Future Shock* (1970), *The Third Wave* (1980) and *Powershift* (1990) – will recognize that there is a heavy emphasis on predicting situations that will probably occur in the short, medium and long terms, internationally, regionally, nationally and locally. Some of these forecast trends and events are more likely than others to have implications for tourism development. Changes in lifestyle, a dramatic reduction in the working week and corresponding increases in leisure time, discretionary income and longevity, for instance, should ensure not only growth in tourism, but also a proliferation in different types of tourism, for the years ahead (Pearce 1989: 290–2).

In the area of tourism development, the analogue to Toffler is most evident among those who regularly engage in futures research. The George Washington University, for example, has for the last two decades organized a series of Delphi studies and colloquia that has sought to gauge the probability and significance of the occurrence of a number of possible global scenarios, along with their differential impacts on tourism. Hawkins (1993) and Ritchie (1993) are at the forefront of such investigations and their work is well known. More germane to the current discussion, however, are their rarer observations of tourism development theory, going as they do under another name.

Ritchie (1993: 214–5) relatedly observes, for instance, that: 'Tourism academics have a major contribution to the development of knowledge in the field . . . in order to advance our understanding of the field of tourism.' For him, the 1990s and beyond will be *radically* different from what has gone before. There will be *rapid* technological innovations, *dramatic* political change, and equally evident sociocultural changes in how people live. 'Tourism as a phenomenon', he says, 'is clearly affected as much, and *perhaps more* by these changes as any other sector' (p. 201, emphasis added). As specific instances of such change, Ritchie highlights the increasing importance of resident-responsive tourism, global lifestyles, demographic shifts and the ever-widening gap between the North and South. Unless collaborative, international, cross-cultural, multidisciplinary, multi-methodological research provides more networked information about these and other developments, there will be parallel serious deficits at the levels of policy, evaluation, management, action and operation.

In a similar vein, Hawkins (1993: 176) maintains that: 'There is a need to address more comprehensively a broad spectrum of tourism issues of major international concerns which are related to the fields of health, energy, transportation, technology, finance, resource conservation, historic preservation and education.' In other words, like Ritchie, Hawkins adopts a domain approach to tourism development, emphasizing that progress in one life sphere impacts on several others (including what people do with their free time). Research is therefore necessary to identify needs in order to operate a successful process model. Interestingly, however, of the nineteen issues and fifty-two research questions he identifies from Delphi responses, only three carry an envisaged theoretical component of 'understanding':

> To what extent can the tourism industry contribute to global restructuring by accepting responsibility for . . . enhancing the understanding of the cultural and natural diversity of the world? (p. 188)
>
> How can new market-oriented economies develop a realistic understanding of the competitive forces operating in the global tourism market? (p. 190)

> How can we improve our understanding of resident perceptions, values and priorities regarding tourism's role in their community? (p. 190)

It can also be argued that such *knowledge*, with its large informational component and stress on need identification, is not quite the same as Jafari's (1989), by now classical, *knowledge-based* platform – i.e. that which emerges after the *pre*-theoretical stages of advocacy, caution and adaptation – but rather an overly descriptive and functionalist approach to knowledge acquisition. Whether such a critique is valid remains unclear from the writings of Ritchie and Hawkins since they never really elaborate on what they intend by *understanding* and whether it is concerned with answering *why*, rather than just *how*, questions. Nevertheless, they are surely correct in maintaining that an appreciation of factors which are interlinked with, and impact on, tourism, is vital *before* a viable agenda for tourism development is put into practice.

THE SIMMELIAN PERSPECTIVE

Formalism in sociology, which is closely identified with the writings of Georg Simmel, adopts a happy midway position between macro theory and micro-theory. However, its usefulness in the current context has more to do with its ability to abstract out the essences of phenomena, those recurring and immutable forms of reality which, though combined with, stand in sharp contrast to, changing content. By definition, these forms transcend both space and time, so much so that they may be regarded as perennial, and therefore *a fortiori* relate to the future, as well as to the past and present. Thus Simmel's essay on the metropolis, for instance, although based on a study of Berlin at the turn of the twentieth century, contains several uncanny intuitive insights which are equally applicable to the New York and Amsterdam of today or to some new conurbation of the future whose plans may not yet be on the drawing board.

As regards tourism research, the most well-known application of Simmel (1950) is Cohen's (1972, 1974, 1979b) work on strangerhood which, when aligned with that of Schutz (1944), lies at the basis of his tourist typologies. Apart from these incisive contributions to our understanding of the tourist, however, and with the possible exception of articles by Brewer (1984), Greenblatt and Gagnon (1983) and Machlis and Burch (1983), for instance, little else has been derived from Simmel and applied to tourism. Yet, arguably, Simmel's analyses of monetary exchange, secrecy, alienation, leadership, rivalry, freedom, friendship, loyalty, authority and the like could provide a useful theoretical framework for studies concerned with tourism development both now and in the years ahead (Dann and Cohen 1991).

Let us quickly identify a couple of recurring features of tourism development where this type of interpretative understanding might prove fruitful. One term that springs to mind is de Weerdt's (1990) expression *vocation touristique* which, although originally applied to French rural tourism, Morris (1995: 183) finds useful in relation to something as different as the Sydney Tower – a shopping telecommunications complex whose audiovisual displays narrate the history of that city's 'tourist vocation' to (white) visitors. However, in the sense adopted by Simmel, vocation (*Beruf*) means something more. It is the response to an invitation to join a group, the *a priori* conditions for which already exist. Thus, among some peoples, the practice of laying a place at table for the unexpected traveller in a way shows that the presence of the wanderer has been anticipated in a culture where hospitality comes spontaneously. As the result of this insight, it might be possible to gain a deeper appreciation of the commoditized aspects of contemporary tourism development by analysing the natural preconditions of the phenomenon which were based on social exchange without reward, where a guest was regarded as someone sacred under the protection of the gods (Przeclawski 1994: 8).

A second example comes from a paper by Chang *et al.* (1996). According to them, site-specific studies fail 'to provide any conceptualization of the processes that underlie urban (heritage) tourism' (p. 287). Future theoretical attempts should therefore examine such global economic forces as deindustrialization, as well as 'the role that local factors can play in mediating top-down processes' (p. 302). However valid this observation may be, arguably a deeper understanding of the attractions of city-based tourism development could have been derived from Simmel's essay on the metropolis. Among the perennial and essential features of this way of life, Simmel identifies dependence on time – so much so that, by way of mental experiment and imaginative reconstruction of alternative reality, he suggests that, if all the clocks stopped, there would be total chaos. (Compare this envisaged situation with the same event occurring in a Third World capital, where it would probably be business as usual with jokes about local attitudes to time.) Other qualities of the city that Simmel outlines include the presence of certain social types – e.g. professional entertainers, strangers, middlemen – the stimulation of nervous energy, persons being treated like a number, and so on, (features which, incidentally, could equally apply to a contemporary theme park). By inversion, therefore, what would be attractive for those interested in the history of a city, would be the *absence* of such postmodern characteristics. When combined with Simmel's thoughts on the fascination of ruins, surely we would have all the main ingredients necessary for understanding urban heritage.

OPEN-ENDED WORK

A third source for identifying future theoretical directions in the study of tourism development (or indeed of any other field of human endeavour) is to examine published research and, in particular, to focus on the beginning and conclusion of its existing accounts. The former, better known as 'a review of the literature', contextualizes the problem, acknowledges the related contributions of others and lays no claim to originality. The latter, also known as 'areas for further investigation', attempts to show how a given study has carried forward the cumulative body of knowledge *and*, in so doing, how it in turn reveals additional avenues for potential inquiry. It is this last feature which is both future-oriented and open-ended, representing therefore a veritable mine of inspiration for those intrigued with possible theoretical developments in the years ahead. Should any of these openings themselves be utilized in a subsequent scholarly offering, they then become part of a new literature review; and so the process continues *ad infinitum*.

However, we have perhaps grown so accustomed to this almost mandatory formatting of research accounts that participation in the exercise can begin to assume many of the qualities of ritual, bearing its own academic reward structure. If this is our jaundiced view in a managed world obsessed with quantification of publications in quality outlets, we may be forgiven. Nevertheless, it would be a pity were this to be our only outlook on scholastic life, since we would surely be missing the main point of sharing in the progress of a discipline or thematic treatment, and the fact that such intellectual development is only possible through the dialectical exchange of ideas.

As far as research on tourism development is concerned, we can usefully explore several ways that future agenda are mapped out in a number of recent papers – some admittedly more fruitful than others – in the hope that they may separately and collectively act as stimuli to an advancement in our understanding of the phenomenon. These various strategies may be arranged according to the degree of confidence they display in a given line of investigation and the willingness to pursue it in collaboration with others. They are ordered under the headings of 'blind alleys', 'self-appropriation', 'reversing conventional wisdom', 'concept stretching', 'scope broadening', 'breaking out of the case', 'resolving paradoxes' and 'establishing new linkages'. The examples they include refer either directly or indirectly to approaches adopted towards tourism development.

Blind alleys

In examining the viability or otherwise of various possible theoretical alternatives, some researchers dismiss at the outset, and in fairly harsh terms, those lines of inquiry which they judge to be less than fruitful. This denigration technique adopts a twofold 'approach-avoidance' strategy. By pointing out blind alleys, it saves others from venturing down theoretical *culs-de-sac*. At the same time, it promotes the chosen path by the rejection of false trails.

Marie-Françoise Lanfant, for example, in attempting to chart a course where identity, change and international tourism are treated coterminously, senses the need to declare straight away that 'tourism as an agent in the destruction of cultures is *a banal theme*' (Lanfant 1995a: 5, emphasis added). At a stroke, therefore, she rejects as less than helpful the stance adopted by many traditional studies of tourism development which view the impact of one (superordinate) external culture on another internal (subordinate variety), in a similar fashion to the collision of billiard balls (cf. Picard 1995). Thus, instead of pursuing a tourism that is 'rooted in an explanatory framework which is characterized by Western ethnocentrism, univocal and reductive' (e.g. in terms of conspicuous leisure, reduction of the working day, democratization of holidays, desire for urban escape, etc.), Lanfant (1995b: 24) argues that it is necessary to view the phenomenon both emically and as an international fact: 'which draws together social groups with contrasting modes of social discourse: developed countries and developing countries; urban populations and rural populations; technological societies and traditional societies; hot societies and cold societies' (Lanfant, 1995b: 30). Hence, the former debate over tradition and modernity in the context of tourism's multidimensional impacts surrenders to an analysis where the pseudo-dichotomy is explored as continuous, and the seeming polar opposites are absorbed into each other (Lanfant 1995b: 36).

Underpinning this argument, however, it should be recognized that Lanfant's theoretical contribution rests on her acknowledged application of a classical insight of Emile Durkheim, that pertaining to social facts. Seen in such a manner, tourism as an international social fact becomes an apparatus of external constraint that can overwhelm tourist choice and the aspirations of destination communities. The agents of this globalization process are clearly the multinational corporations of the so-called 'New Economic Order', which often work in league with various UN and other world bodies by transcending the boundaries of nation states, in much the same way as tourism itself does (Lanfant 1980). Regarded in this light, tourism development becomes part of a much wider capitalistic process that places limits on the exercise of individual freedom.

It should also be noted that, in their enthusiasm to discourage certain theoretical avenues of investigation, the language of researchers can occasionally border on the extreme. Thus, Salamone (1997: 306–7), for instance, dismisses a superorganic approach to so-called *tourist culture* as reducible to a 'solipsistic nihilistic relativism', and questions regarding a culture's authenticity as degenerating into 'mere platonic quibbles for those with political axes to grind' (Salamone 1997: 319).

If one goes beneath the rhetoric, however, the message for tourism development theory becomes evident. Grand theory (*à la* Parsons or Marx, for example), is only useful in conceptualizing tourism systemically or functionally. It does so at the risk of overlooking all the micro-developments associated with community-based approaches (Murphy 1985).

Of course, dismissal of adversaries in order to reinforce a given theoretical position is not a new academic technique. It stands at the basis of Socratic dialogue and the medieval scholastic disputation, and indeed runs throughout all the major works of Thomas Aquinas. Its application to tourism development, however, can be both novel and rewarding, particularly where themes are of a controversial nature, for example, tourism's alleged potential to foster world peace (Dann 1988; Crick 1989; Smith 1998).

Self-appropriation

Sometimes the pursuit of a specific theoretical line of inquiry can be quite a personal matter, so much so that authors prefer to handle the task entirely by themselves, rather than leaving it for others (to ruin). Nowhere is this attitude more evident than in the relatively recent anthropological concern with the role of the observer in studies of Third World tourism (cf. Crick 1985, 1995). In this regard, Bruner (1995: 238), for example, writes: 'Tourism is primarily visual, ethnography verbal. Tourists surrender, ethnographers struggle. Possibly even more important are two points about which I am still gathering data and *am not prepared to discuss at this time*, but I will mention them' (emphasis added). Having thus whetted his audience's appetite, Bruner then discloses the nature of the unfinished business – the disavowal of moral and political responsibility by tourists and its acceptance by ethnographers, and the impossibility of differentiating in the culture of mature destinations between the touristic and the ethnographic.

The self-appropriation of agenda is made all the more dramatic by the use of the first person singular – a feature more noticeable in the anthropological treatment of tourism development than in any other social scientific discipline.

Reversing conventional wisdom

Another way of establishing an exciting theoretical agenda is to debunk popular assumptions and, via 'sociological (and other) imagination' (Mills 1959), to reverse conventional wisdom. This approach does not completely jettison what has gone before. It merely alters the direction of dependency, turning it on its head.

Thus, Kruhse-Mount Burton (1995), for example, examines the commonly held view that tourism development and prostitution go hand in hand, so much so that it is widely believed that the former causes the latter. Untrue, she argues; prostitution existed hundreds of years before the advent, let alone the expansion, of tourism. Another more satisfactory explanation therefore has to be sought as to why exactly men seek female extramarital companionship, especially in cultures different from their own (the case of Australian sex tourism in the Far East, for instance). The answer, she suggests, may rather lie in the motivational push-pull framework that is still frequently used in an analysis of contemporary *non-sex* tourism. More specifically, by emphasizing the nostalgia factor in tourist motivation, we can begin to understand how such men wish to turn the clock back to the period when they were little boys by visiting areas whose populations are promotionally captured in a similar time warp.

The reversal of conventional wisdom is often also accompanied by startling revelations. Smith (1998: 206), for instance, clinically observes that 'if there is a war, there is tourism'. Indeed, warming to her theme, she declares: 'Despite the horrors and destruction (and also *because* of them), the memorabilia of warfare and allied products . . . probably constitute the *largest* single category of tourist attractions in the world' (Smith 1998: 205). Here emphasis has been added to show the intended causal link and to highlight the up-ending of the received knowledge that tourism thrives in and contributes to an atmosphere of peace. Not necessarily so, says Smith. Not only do the fastest rates of tourism growth occur in post-war periods (*post hoc, ergo propter hoc*), but there are now new forms of tourism (emotional, military and political) which cater on a massive scale to such demand. Hence the increasing popularity of medieval jousting contests, re-enactments of famous battles, cemetery tours, holocaust camps and sites of atomic disaster. Whether or not we like her focus, Smith is surely correct in concluding that these dark features of contemporary tourism can and should be included under tourism development.

Concept stretching

Van der Borg *et al.* (1996: 308) have recently stated that 'sustainability has become a central issue in much of today's tourism development literature. However the application of the concept of sustainable tourism development has largely been limited to non-urban or rural areas'. They thus extend a theoretical construct from one setting in order to explore its viability in another – in this instance, the growth of tourism in European heritage cities. In so doing, of course, they should also be required to examine two controversial expressions: 'sustainable tourism' and 'sustainable development' (Butler 1997; Wall 1997), terms which themselves are the result of concept stretching.

Arguing from the opposite end of the continuum, as it were, Oppermann (1996: 86–7) points out that, although there has been a long tradition of rural tourism, the phenomenon has not been matched by an equally lengthy research tradition. In fact, conceptually, he says, there is not even a commonly accepted definition as to what rural tourism is. He consequently maintains that it is necessary to investigate the interrelationships between urban centres and rural areas and vice versa. Moreover,

> while most research has focused on either hosts or guests and a few on both, studies on host – guest interactions and interrelations are rare. Thus another important question is whether guests contribute to hosts' innovative potential by transferring ideas from urban to rural areas.

Scope broadening

Closely allied to 'concept stretching' is the perceived need to go beyond operationalized ideas and variables to broaden the whole scope of a research project at a later stage. A familiar clue in the literature that such an agenda has been postponed is some such expression as 'although outside the scope of this paper . . . it would be of considerable interest to know more about . . .' (e.g. Timothy and Wall 1997: 336).

In this regard, Parrinello (1993: 242), for instance, while acknowledging the importance of advertising in an increasingly symbolic discourse of tourism promotion within the surrounding postindustrial ethos of many tourism-generating societies, shows how investigations can go further than simply restricting themselves to analyses of single content messages by contextualizing them within another facet of the postmodern condition – that of commoditization. Consequently, she recommends extending the analysis to those cases where luxury consumer products (e.g. cars, cameras, perfume) are simultaneously promoted with exotic destinations. Taking a cue from her suggestion, this writer has recently discovered a number of related marketing techniques featured in a leading travel magazine (Dann forthcoming), thereby emphasizing the growing importance of imagery in studies of tourism development (Dann and Potter 1994; Gartner 1997).

Another example of this sort of broadening strategy may be found in Hughes' (1997) recent investigation of resorts that cater to male homosexuals. By extending the traditional approach to tourist identity, as evidenced in the work of MacCannell (1989) and Bruner (1991), for instance, to include a gendered dimension, he is able to claim: 'the holiday, nonetheless, is likely to make a significant contribution to the creation and validation of identity for many gay men', before tellingly adding as future agenda: 'the nature and dimension of this contribution remain to be examined further' (Hughes 1997: 7).

By way of complement, Wall (1997) identifies feminist studies as constituting a recent and important, if not sometimes controversial, approach to theories of tourism development. In a similar fashion to Hughes, he laments their relative dearth of depth.

Breaking out of the case

In an article about using Santa Claus as a means towards tourism development in Lapland, Pretes (1995) is very much aware that he needs to break out of the case of an Arctic village near Rovaniemi. By investigating why tourists should wish to visit this specific, and otherwise unremarkable, site (thereby yielding knowledge useful for local tourism planners and marketers), he quickly realizes that the answers to his research question (in terms of a romantic nostalgia for childhood) may also provide a degree of generalization that extends beyond a particular locale in Finland to similar themed hyperreal destinations elsewhere (e.g. Disney World). Why? Because, in his words: 'Perhaps *more importantly*, an *understanding* of tourist behavior might illuminate a *deeper* sociological question: what *motivates* tourists to visit overtly contrived attractions?' (Pretes 1995: 2; emphasis added). Thus, according to Pretes, and indeed many others, it is theory ('understanding') that makes transcendence of the ideographic and attainment of the nomothetic possible. Furthermore, this explanatory task is facilitated by confronting what is surely the most theoretical issue of all – motivation – which, as Max Weber himself has pointed out on several occasions, lies at the core of causal interpretation (Dann *et al.* 1988).

In relation to tourism development, however, Pretes' agenda raise another important theoretical concern, namely, the viability, and perhaps desirability, of substituting contrived and artificial for genuine and natural attractions on the grounds of greater environmental sustainability and lesser invasion of resident privacy (Buck 1977). By necessarily locking into the wider theoretical debates of authenticity (e.g. Cohen 1979b, 1995; Boorstin 1987; MacCannell 1989) and placelessness (Relph 1983), and their respective associations with tourism's differential impacts, he is therefore once again able to break out of the single case.

'Resolving paradoxes'

One of the hallmarks of theory, according to Berger's (1963) classical account, is that it should be sociologically problematic, which is to say that it should provide an understanding of the familiar (which we often, erroneously, believe that we understand already). This approach of Berger's stands in sharp contrast to the position which considers reality as a series of social problems. Thus, for him, before we study the pathological or deviant, we should be trying to come to terms with the normal – before we examine divorce, for example, we need to have an understanding of marriage. For this writer, it would seem to follow from the above that strategies which increase awareness of the problematic also enhance theoretical awareness. One such strategy is developing a consciousness for paradox.

In tourism development research, there are several instances where practitioners have made their audience aware of the existence of paradoxes, and, arguably, at the same time, the theoretical content of their accounts. One example is Wilson's (1997) work in Goa. Why is it, he asks, that so many tourists are so enthusiastic about this Indian island destination when there are several 'aspects of the tourist experience, which, if found in Europe or other developed countries, would constitute serious grounds for complaint?' (p. 53). Again, 'why has tourism development been condemned so strongly in Goa in spite of the fact that its economic benefits are spread so widely through the local population, many of whom seem to welcome tourism and tourists?' (p. 57). Although not providing all the answers to these questions, and to that extent opening up the discussion to further research, Wilson nevertheless speculates about some solutions. As regards the first paradox, he indicates that a certain trade-off exists in the tourist's mind between lack of hygiene, high prices and drugs, on the one hand, and the good weather, friendly people, feelings of safety and relaxation, on the other – a sort of mental cost-benefit analysis borrowed from economic theory and translated into psycho-social attitudes.

The second paradox may not be due so much to the traditional tendency of attributing to tourism development the negative consequences associated with coterminous phenomena (e.g. urbanization and industrialization), but rather due to international tourism's being incorrectly assigned the blame for expansion in domestic tourism (which could equally or even more be at fault). It is this last possibility which leads Wilson to acknowledge that it is a weakness of his own research that he did not address this issue sufficiently (p. 65), and hence by implication, that this area too is necessarily conducive to future agenda.

Establishing new linkages

One way of advancing the theoretical cause in tourism development is to take two or more domains and combine them. Hence, Xiao (1997: 358), in a study of China, contends that: 'despite the multifaceted nature of research on leisure and tourism, fundamental questions concerning their interrelationship still remain unsolved'. Consequently, he proposes that, instead of assuming that tourism development is profit-oriented (for visitors) and leisure development is welfare-oriented (for residents), one could consider a leisure experience as also being a tourism experience. By adopting such an approach, it is thus possible to examine the conditions under which tourism becomes leisure, the contributions of local communities to leisure and tourism, and the social factors which seem to integrate tourism and leisure.

Similarly, Pearce (1997), in his analysis of tourism development in Spain, maintains that political science and development theory need to be integrated in order to obtain a more complete picture of what is taking place. 'Both these processes,' he states, 'political transformation and tourism development have attracted considerable attention, but relatively little work has focused on the *links between them*' (Pearce 1997: 157, emphasis added). Interestingly, whereas he claims that his own study contributes towards 'filling this gap', he nevertheless regards its contribution as only a first step. Far from considering it as closure of the discussion, he readily acknowledges on three other separate occasions (pp. 168, 173, 175) that his own offering opens itself to future research. The last of these admissions is perhaps the most significant since it introduces the need for theory. In his words: '*Further work* on tourism organizations is now needed, *to increase knowledge* of these important players and *improve understanding* of other activities such as these which are commonly only seen from a technical viewpoint' (Pearce 1997: 175, emphasis added).

Finally, a most ambitious attempt to seek linkages through multidisciplinary treatment is evident in a recent paper by Bramwell and Rawding (1996). Here the authors deal with the vital question of tourism promotion, and, more specifically, the ways that old industrial cities give themselves a new lease of life through tourism development. Since the focus is on imagery, and imagery itself is open to multidisciplinary analysis, they seek simultaneous insights from geography, marketing and critical sociology. Germane to the current presentation is the future theoretical orientation of their study, since Bramwell and Rawding clearly identify what has been explained by each discipline and what still remains to be understood. The latter consideration in turn breaks new ground as soon as it leads to the larger (relatively unexamined) questions of inequalities of wealth and power, the distribution of tourism's benefits, community participation in self-promotion, and the development of a sense of place. Their work constitutes a prime exemplar of concern with tourism theory which is both multidisciplinary and future-oriented, and, as such, could well stand as a model for others in investigating tourism development.

CONCLUSION

This account has examined three ways of elaborating a future research agenda in tourism development studies. Although the strategies are initially quite different, interestingly they converge in a perceived need for multidisciplinarity.

The futures approach adopted by Hawkins and Ritchie, for example, reached a point where they called for collaborative, international, cross-cultural, multidisciplinary and multi-methodological research in tourism (Ritchie 1993: 202–3). The Simmelian perspective, although less obvious in this regard, can also be considered in terms other than merely sociological.

An analysis of strangerhood, for example, could benefit from additional insights provided by psychology, geography, history, psychiatry and political science. So too could the allied notions of belonging, place and placelessness, alienation and vocation, all crucial for an understanding of tourism development in the past, present and in the years ahead. At the moment we simply do not have studies which explore these perennial realities theoretically from a combination of disciplines and paradigms.

By examining black holes in the literature, we similarly arrive at a juncture, epitomized by the work of Bramwell and Rawding (1996), where several disciplines are simultaneously co-opted in order to yield a combined understanding that opens itself to critical issues that for too long have remained largely unexplored.

Crouch (1991: 67–9) states the situation somewhat differently, albeit in complementary terms. Knowledge building in tourism, he says, is rather like the construction of a physical edifice. In the beginning we may only have a vague idea of the sort of building we require, and this vagueness is often reflected in the site plan that is prepared. As construction proceeds, inevitably some of the foundations will be shaky, resting, as they do, on theoretically swampy ground, where even continued excavation fails to find solid bed-rock. Other foundations prove far sounder, and these can later act as a basis for further extension. For Crouch, designing the site plan need not be such a precarious business, particularly if we pay heed to construction work that has gone on before, as evidenced principally and analogously through reviews of the literature.

This chapter adopts a similar view. While agreeing with Pearce (1993: 1) that 'the tourism literature has never been rent by widespread ongoing debates about theoretical and methodological issues', it nevertheless maintains that such a situation is ripe for change. It is simply a question of compiling a lengthy list of areas for future research and then prioritizing the issues in the form of viable agenda. Just how this undertaking should be carried out can be seen by a brief return to Hawkins' (1993) process model. After issues for tomorrow's tourism development are identified via the Delphi technique, research questions are subsequently deduced. Thus, for instance, one of the nineteen tourism issues to emerge from the brainstorming of experts is the need for greater 'resident responsive' tourism, that is to say, a more democratic participation in tourism decision making by grassroot members of a destination society. Hawkins (p. 190) indicates that this issue in turn leads to three research questions:

- the identification of measures which ensure that tourism development is in harmony with the sociocultural, ecological and heritage goals of the local community, along with any other related values and aspirations;

- the search for creative approaches towards fostering citizen participation in the economic benefits of tourism development;
- the furtherance of understanding of resident perceptions, values and priorities regarding tourism's role in the community.

Now, of these three research questions, and as noted earlier, only the last is strictly theoretical since it seeks understanding. However, with a little imagination, the first two may also be seen to generate theoretical concerns. The use of the word 'harmony', for example, implies the critical application of some sort of consensus perspective, just as its opposite, 'conflict', suggests a reworking of the trickle-down dependency paradigm. The notion of a 'greater sharing in decision making' by those directly affected by tourism development, but who may lack the necessary expertise for such deliberation, may also be highly problematic and, to that extent, theoretical. Exchange theory could arguably fill the void.

The same framework can also be fitted to many of Hawkins' other key issues, for instance:

- recognition of the finite limitations to tourism development (in contrast to the infinite aspirations exploited by tourism promotion);
- the growing importance of environmental awareness (and the clash with the politics and practice of alternative vested interests in green sites);
- the increasing cost of capital (and the competition of rival concerns e.g. health and education);
- rapid advances in technology (and, with the advent of 'post-tourism', a reduction in the need for conventional tourism development);
- globalization (and the necessity for proactive responses from the tourism industry);
- the tension between centre and periphery in tourism development (and the implications for national versus regional initiatives);
- variation in tourism demand (as dependent on changes in lifestyle).

All of the foregoing have theoretical components but very little seems to have been explored along those lines.

Similarly, many areas for further research, so easily identified and extracted, both *a priori* from the Simmelian perspective and *a posteriori* from the open-ended approach, appear to receive scant theoretical attention in the literature. Why should this vital ingredient of research be missing? According to other contributors to this volume (Pearce and Moscardo, Chapter 3), and the previously cited Peter Berger (1963), a theoretical focus can only be generated by asking the right questions. Unless issues are problematized – unless we acknowledge that our understanding is incomplete – we will never adequately address issues of tourism development in the present, let alone in the proximate future.

References

Berger, P. (1963) *Invitation to Sociology*, New York: Doubleday.

Boorstin, D. (1987) *The Image. A Guide to Pseudo Events in America*, 25th anniversary edition, New York: Atheneum.

Bramwell, B. and Rawding, L. (1996) 'Tourism marketing images of industrial cities', *Annals of Tourism Research* 23 (1): 201–21.

Brewer, J. (1984) 'Tourism and ethnic stereotypes: variations in a Mexican town', *Annals of Tourism Research* 11 (3): 487–501.

Bruner, E. (1991) 'Transformation of self in tourism', *Annals of Tourism Research* 18 (2): 238–50.

—— (1995) 'The ethnographer/tourist in Indonesia', pp. 224–41 in M.-F. Lanfant, J. Allcock and E. Bruner (eds) *International Tourism. Identity and Change*, London: Sage.

Buck, R. (1977) 'The ubiquitous tourist brochure. Explorations in its intended and unintended use', *Annals of Tourism Research* 4 (4): 195–207.

Butler, R. (1997) 'Modelling tourism development: evolution, growth and decline', pp. 109–25 in S. Wahab and J. Pigram (eds) *Tourism, Development and Growth. The Challenge of Sustainability*, London: Routledge.

Chang, T., Milne, S., Fallon, D., and Pohlmann, C. (1996) 'Urban heritage tourism. The global–local nexus', *Annals of Tourism Research* 23 (2): 284–305.

Cohen, E. (1972) 'Toward a sociology of international tourism', *Social Research* 39 (1): 164–82.

—— (1974) 'Who is a tourist? A conceptual clarification', *Sociological Review* 22 (4): 527–55.

—— (1979a) 'Rethinking the sociology of tourism', *Annals of Tourism Research* 6 (1): 18–35.

—— (1979b) 'A phenomenology of tourist experiences', *Sociology* 13 (2): 179–201.

—— (1995) 'Contemporary tourism – trends and challenges: sustainable authenticity or contrived post-modernity?', pp. 12–29 in R. Butler and D. Pearce (eds) *Change in Tourism. People, Places, Processes*, London: Routledge.

Crick, M. (1985) '"Tracing" the anthropological self: quizzical reflections on fieldwork, tourism and the ludic', *Social Analysis* 17: 71–92.

—— (1989) 'Representations of international tourism in the social sciences. Sun, sex, sights, savings and servility', *Annual Review of Anthropology* 18: 307–44.

—— (1995) 'The anthropologist as tourist: an identity in question', pp. 205–23 in M.-F. Lanfant, J. Allcock and E. Bruner (eds) *International Tourism. Identity and Change*, London: Sage.

Crouch, G. (1991) 'Building foundations in tourism research', pp. 67–75 in R. Bratton, F. Go and J. Ritchie (eds) *New Horizons in Tourism and Hospitality Education, Training and Research*, Calgary: World Tourism Education and Research Centre, University of Calgary.

Dann, G. (1988) 'Tourism, peace and the classical disputation', pp. 25–33 in L. D'Amore and J. Jafari (eds) *Tourism. A Vital Force for Peace*, Montreal: First Global Conference.

—— (1998) 'The other side of tourism: punching around in the dark', public

lecture given to postgraduate students and staff of the Scottish Hotel School, University of Strathclyde, 12 March.
—— (forthcoming) 'The pomo promo of tourism', *Tourism, Culture and Communication*, 1, 1.
Dann, G. and Cohen, E. (1991) 'Sociology and tourism', *Annals of Tourism Research* 18 (1): 155–69.
Dann, G., Nash, D., and Pearce, P. (1988) 'Methodology in tourism research', *Annals of Tourism Research* 15 (1): 1–28.
Dann, G. and Potter, R. (1994) 'Tourism and postmodernity in a Caribbean setting', *Cahiers du Tourisme*, serie C, no. 185.
de Weerdt, J. (1990) 'L'espace rural français: vocation touristique ou processus de touristification?', paper presented to the working group on international tourism of the International Sociological Association, Madrid, July.
Foley, M. and Lennon, J. (1996) 'JFK and dark tourism: a fascination with assassination', *International Journal of Heritage Studies* 2 (4): 198–211.
Gartner, W. (1997) 'Image and sustainable tourism systems', pp. 179–96 in S. Wahab and J. Pigram (eds) *Tourism, Development and Growth. The Challenge of Sustainability*, London: Routledge.
Greenblatt, C. and Gagnon, J. (1983) 'Temporary strangers: travel and tourism from a sociological perspective', *Sociological Perspectives* 26 (1): 89–110.
Hawkins, D. (1993) 'Global assessment of tourism policy: a process model', pp. 175–200 in D. Pearce and R. Butler (eds) *Tourism Research. Critiques and Challenges,* London: Routledge.
Hughes, H. (1997) 'Holidays and homosexual identity', *Tourism Management* 18 (1): 3–7.
Jafari, J. (1989) 'Sociocultural dimensions of tourism. An English language literature review', pp. 17–60 in J. Bystrzanowski (ed.) *Tourism as a Factor of Change. A Sociocultural Study*, Vienna: Vienna Centre.
Kruhse Mount-Burton, S. (1995) 'Sex tourism and traditional Australian male identity', pp. 192–204 in M.-F. Lanfant, J. Allcock and E. Bruner (eds) *International Tourism. Identity and Change*, London: Sage.
Lanfant, M.-F. (1980) 'Tourism in the process of internationalization', *International Social Science Journal* XXXII, 1: 14–43.
—— (1995a) 'Introduction', pp. 1–23 in M.-F. Lanfant, J. Allcock and E. Bruner (eds) *International Tourism. Identity and Change*, London: Sage.
—— (1995b) 'International tourism, internationalization and the challenge to identity', pp. 24–43 in M.-F. Lanfant, J. Allcock and E. Bruner (eds) *International Tourism. Identity and Change*, London: Sage.
MacCannell, D. (1989) *The Tourist. A New Theory of the Leisure Class*, 2nd edn, New York: Schocken Books.
Machlis, G. and Burch, W. (1983) 'Relations between strangers: cycles of structure and meaning in tourist systems', *Sociological Review* 31 (1): 666–92.
Mills, C. (1959) *The Sociological Imagination*, New York: Grove Press.
Morris, M. (1995) 'Life as a tourist object in Australia', pp. 177–91 in M.-F. Lanfant, J. Allcock and E. Bruner (eds) *International Tourism. Identity and Change*, London: Sage.
Murphy, P. (1985) *Tourism: A Community Approach*, London: Methuen.

Oppermann, M. (1996) 'Rural tourism in southern Germany', *Annals of Tourism Research* 23 (1): 86–102.

Parrinello, G.-L. (1993) 'Motivation and anticipation in post-industrial tourism', *Annals of Tourism Research* 20 (2): 233–49.

Pearce, D. (1989) *Tourist Development*, 2nd edn, Harlow: Longman.

—— (1993) 'Introduction', pp. 1–8, in D. Pearce and R. Butler (eds) *Tourism Research. Critiques and Challenges*, London: Routledge.

—— (1997) 'Tourism and the autonomous communities in Spain', *Annals of Tourism Research* 24 (1): 156–77.

Pearce, P. (1982) *The Social Psychology of Tourist Behaviour*, Oxford: Pergamon.

Picard, M. (1995) 'Cultural heritage and tourist capital: cultural tourism in Bali', pp. 44–66 in M.-F. Lanfant, J. Allcock and E. Bruner (eds) *International Tourism. Identity and Change*, London: Sage.

Pretes, M. (1995) 'Postmodern tourism. The Santa Claus industry', *Annals of Tourism Research* 22 (1): 1–15.

Przeclawski, K. (1994) *Tourism and the Contemporary World*, Warsaw: Institute of Social Prevention and Readaptation, Centre for Social Problems of Education, University of Warsaw.

Relph. E. (1983) *Place and Placelessness*, London: Pion.

Ritchie, J. (1993) 'Tourism research. Policy and managerial priorities for the 1990s and beyond', pp. 201–16 in D. Pearce and R. Butler (eds) *Tourism Research. Critiques and Challenges*, London: Routledge.

Salamone, F. (1997) 'Authenticity in tourism. The San Angel inns', *Annals of Tourism Research* 24 (2): 305–21.

Schutz, A. (1944) 'The stranger: an essay in social psychology', *American Journal of Sociology* 49 (6): 495–507.

Seaton, A. (1996) 'Guided by the dark: from thanatopsis to thanatourism', *International Journal of Heritage Studies* 2 (4): 234–44.

Simmel, G. (1950) 'The stranger', pp. 401–8 in K. Wolff (ed.) *The Sociology of Georg Simmel*, Glencoe: Free Press.

Smith, V. (1998) 'War and tourism: an American ethnography', *Annals of Tourism Research* 25 (1): 202–27.

Timothy, D. and Wall, G. (1997) 'Selling to tourists. Indonesian street vendors', *Annals of Tourism Research* 24 (2): 322–40.

Toffler, A. (1970) *Future Shock*, New York: Random House.

—— (1980) *The Third Wave*, New York: William Morrow.

—— (1990) *Powershift*, New York: Bantam Books.

van der Borg, J., Costa, P., and Gotti, G. (1996) 'Tourism in European heritage cities', *Annals of Tourism Research* 23 (2): 306–21.

Wall, G. (1997) 'Sustainable tourism – unsustainable development', pp. 33–49 in S. Wahab and J. Pigram (eds) *Tourism, Development and Growth. The Challenge of Sustainability*, London: Routledge.

Wilson, D. (1997) 'Paradoxes of tourism in Goa', *Annals of Tourism Research* 24 (1): 52–75.

Xiao, H. (1997) 'Tourism and leisure in China. A tale of two cities', *Annals of Tourism Research* 24 (2): 357–70.

3

TOURISM COMMUNITY ANALYSIS

Asking the right questions

Philip L. Pearce and Gianna Moscardo

INTRODUCTION

This chapter addresses two interwoven themes. The first theme is concerned with how researchers generate their studies; that is, how they frame and ask questions. This theme is illustrated in the chapter by a second strand of attention: the direction, shape and future possibilities for tourism researchers in the area of tourism community relationships. The chapter will use question-asking analysis to provide some insights into tourism community research and might appeal to two audiences, those concerned with tourism questions in general and those with a specific interest in tourism community impacts.

Question asking in context

As tourism has developed and research in the field has become more prominent and the need for it better articulated and recognized, numerous attempts to identify tourism research agenda have been undertaken (Ritchie and Goeldner 1994). Some of these agenda have been the efforts of government, others have been industry exercises, while yet others have had an academic orientation or are the results of co-operative endeavours (Faulkner *et al.* 1995). The area of interest of the present paper, tourism community relationships, is frequently cited in such research planning documents and is often given priority status on the list of global, national and local tourism research agendas. For example, the World Tourism Leaders' Meeting on the Social Impacts of Tourism suggests that countries should, as a first priority, 'support greater involvement of communities in the planning implementation monitoring and evaluation processes of tourism policies programs and projects' (WTO 1997a). Similarly, the Asia

Pacific Ministers' Conference on Tourism and the Environment suggested that tourism should 'foster local community involvement and integrated tourism planning for sustainability' (WTO 1997b).

Ritchie (1993), reviewing research agenda items for global tourism policy resulting from discussion groups with ninety senior tourism personnel including government, industry and academics, from twenty-six countries noted a leading agenda item as providing resident responsive tourism. Further, Ritchie identified fifteen research needs germane to this area (Pearce, Moscardo and Ross 1996: 10).

As the study of tourism develops, it is clear that setting tourism research agenda is an engaging and frequently repeated activity. Issues such as sustainability and quality are common themes underlying these agenda. One of the challenges for researchers examining these research agenda is how to proceed from these generic directions to the development of insightful and incisive research. It is proposed in this chapter that some careful attention to the topic of how to ask questions and what kinds of questions to ask represents a fruitful pathway from the numerous research agenda to actual studies.

A question-asking analysis

Three issues can be identified as fundamental to the understanding of question asking. The first concern is the importance of the level and depth of the question. Undoubtedly many researchers have shared the personal experience of being involved in a complex conversation where fellow researchers were asking what appeared to be an obvious question. For example, a group of philosophers discussing the question 'how do I know you are here?' might generate a simple and naive positivist reply from an outsider – 'because I can see you'. To the philosophers this is a comical and inadequate explanation for a substantial problem in ontology. In tourism, questions such as why do people travel and does tourism have an impact on the environment may seem to be reasonable questions. On closer analysis, however, these questions also require substantial elaboration and refinement. Clearly the level and depth at which the question is framed and the kind of answer which can be expected is an important consideration in question asking.

A second concern in question analysis is the need for the question to have clear focus. Lord Florey, the Nobel prize-winning chemist who pioneered the development of penicillin, was by most accounts a demanding and stimulating supervisor and teacher (Bickel 1983). He was fond of wandering around his Oxford laboratories and assailing his research workers with a simple question: 'Tell me, as simply as you can, what is the question you are working on?' This kind of emphasis on keeping researchers focused on the core question driving their research is valuable, to avoid being side-tracked by fascinating yet tangential issues. The need for tourism researchers to know exactly what they are trying to determine in their studies is an important component of question asking.

The third issue in question analysis is the challenge of relevance. Undoubtedly many tourism researchers have experienced the response of industry personnel who challenge academic analysts with the query 'So what?'. This challenge can be met on a number of levels, as our studies into tourism may have short-term and practical consequences, they may have a long-term, generative application, or they may be inherently fascinating as a study of the changing nature of contemporary society. The challenge of relevance really poses the questions: relevant to whom, in what time frame and for what purpose? Irrespective of our approach to answering the issue of relevance it remains an important dimension in question-asking analysis.

One of the concerns of this chapter is to understand the question asking process so some insight can be generated into the way our tourism study areas are developing. Four processes can be identified which take researchers from a general research area to a research question. First, a researcher can extend an existing theory or revise and modify an aspect of an existing framework. The question that they are asking may therefore be either directly or indirectly theoretically driven. A second process which stimulates new questions derives from a technological or industry innovation. For example, a number of questions are being raised by Internet communication practices in tourism marketing and so questions about how the new practice fits in to the existing pattern of the tourism system can be generated (Schonland and Williams 1996; Walle 1996). For many tourist researchers one also suspects there is often an autobiographical element involved in motivating the individual to develop their research question. It might be an acute observation, a personal experience or a long held fascination with a topic which takes them from the research area to a research question. Such autobiographically driven questions have the power to be both meaningful to the individual and to a wider community. And a fourth process is where an industry, individual, business or organization identifies a problem which they feel needs action and solution and the genesis of a research question lies in this activity. Irrespective of the process by which people move from a research area to a research question the good research question is going to have to be answered within the resources of the researcher and subject to good scientific practice in terms of rigour, replication and refutation (cf. Ryan 1995).

Three broad categories of questions can be identified. There are questions of description such as how many, what is, how long, how far and how much? There are questions of difference such as which is bigger, which is more important, which is preferred; and there are questions of relationship, causation and pattern such as what correlates with what, what is linked to what, what depends on what, and what summarizes the pattern in the total array of information.

In summary, for tourism researchers analysing the process of question asking it is important to consider the level and depth of the question, the focus of the question, and its relevance for diverse audiences. Further, the genesis of the question may vary and can include whether or not the question is a theoretical extension, whether or not the question represents an industry innovation, relates to a personal biography or derives from an industry problem or action. Finally the category of questions we are concerned with may also vary in terms of description, difference, relationship or pattern. Armed with this framework, we will now examine the study of tourism community relationships as a case study of question asking.

TOURISM COMMUNITY STUDIES

Tourism community relationships represent a research agenda of some significance in the total picture of studies on the development of tourism. The works of Hawkins (1993) and Ritchie (1993), of Krippendorf (1987) and Murphy (1981, 1985, 1998), have all provided a basis for highlighting the centrality of tourism community relationships in the future of tourism. In an identification of major issues shaping global tourism policy Hawkins (1993) lists nineteen issues, five of which are centrally linked to the social and community impacts of tourism. Similarly, Ritchie (1993) offers a research agenda to encourage and facilitate resident-responsive tourism which he has also identified as a core driving force of tourism for the new millennium. Murphy's work on a community-driven tourism planning approach has an explicit recognition that communities must be assessed in terms of their perception, preferences and priorities for tourism planning and development while Krippendorf directly argues for and describes new approaches to planning for tourism involving residents. The tourism community relationships issue is central to arguments about sustainable tourism and despite the dominance of a biophysical language to describe the resources of tourism in the sustainability debate, the original conceptions of this term were all closely aligned to a concern with the well-being of communities (Pearce 1995).

Tourism researchers have made some advances and have been productive in the area of understanding tourism community relationships. There are a substantial number of published studies, particularly in the last three decades. It is the purpose of this chapter to review what these studies have been saying; to consider the kinds of questions that the researchers have been asking in this branch of tourism analysis and then to further contemplate how new questions might be generated which add to the insights of the existing work. It is important that these general remarks be buttressed by sound evidence. Using an exacting sampling procedure, 262 references were drawn from the existing tourism literature published in English on tourism community relationships and impacts. A smaller set of 110 publications was then further identified which described research on or theories about either tourism's social and cultural impacts or community attitudes towards tourism. Commentaries, reviews and planning literature make up the rest of the total sample. The source of this material lies in the *Annals of Tourism Research* and the *Journal of Travel Research* from the first issue until the mid 1990s plus a less intensive search of twelve major journals in tourism. In addition, *Current Contents* for the Social and Behavioural Sciences was investigated and cross-checking with the reference lists of the previously compiled literature was also undertaken. The results of this search process are listed in two ways. One group of research studies can be classified as 'case studies' involving ethnographic and qualitative approaches to tourism impacts (Table 3.1). A second set of studies based on survey work can also be identified (Table 3.2). While these tables are not meant to be definitive or all inclusive they cover a very substantial body of the refereed research in this field. The existing literature presented in these tables can be summarized as consisting of two major methodological traditions and two associated conceptual approaches.

The first methodological tradition is that of the ethnographic case study; most of these studies have been conducted in developing countries and take a cautionary perspective emphasizing the negative consequences of tourism. Several stage models form the major conceptual background to this kind of research. The stage models, such as those of Doxey (1975) and Butler (1980), while not always developed directly to tackle tourism community relationships, have been used as a broad framework within which specific case studies have been interpreted. These stage models suggest that as tourism development increases in a destination most communities suffer from a variety of negative impacts and almost inevitably become at least concerned about and sometimes hostile to tourism. The ethnographic tradition was supplemented and to some extent supplanted by a second methodological tradition, social survey research.

The social survey work typically attempts to detail and examine host community attitudes towards tourism and its impacts. Many different variables have been studied in numerous locations but few consistent patterns or relationships have been uncovered (Table 3.3). In general there is some limited support for a relationship between increased tourism development and greater awareness of tourism impacts. Some evidence is also available to suggest the existence of equity or social exchange processes which constitute the second major conceptual approach to the study of tourism community relationships. Such an approach proposes that individuals balance the costs and benefits of tourism and their support for tourism depends on the outcome of this cost–benefit equation. Unfortunately both of the conceptual models which exist are too simple to explain the complex pattern of results that have been reported and, whether we are

Table 3.1 Case studies, ethnographic and qualitative approaches to tourism impacts

Akau'ola, L., L. 'Ilaiu, and A. Samate
1980 The social and cultural impact of tourism in Tonga. In *Pacific Tourism as Islanders See It*, South Pacific Social Sciences Association and The Institute of Pacific Studies. Fiji: Institute of Pacific Studies of the University of the South Pacific and the South Pacific Social Sciences Association.

Andronicou, A.
1979 Tourism in Cyprus. In *Tourism. Passport to Development? Perspectives on the Social and Cultural Effects of Tourism in Developing Countries*, E. de Kadt (ed.). New York: Oxford University Press.

Awekotuku, N.T.
1977 A century of tourism. In *The Social and Economic Impact of Tourism on Pacific Communities*, B.H. Farrell (ed.). Santa Cruz: Centre for South Pacific Studies, University of California.

Brougham, J.E., and R.W. Butler
1981 A segmentation analysis of resident attitudes to the social impact of tourism. *Annals of Tourism Research,* 8, 569–90.

Boissevain, J.
1979 The impact of tourism on a dependent island. Gozo, Malta. *Annals of Tourism Research* 6(1): 76–90.

Cohen, E.
1982a Thai girls and Farang men. The edge of ambiguity. *Annals of Tourism Research* 9(3):403–28.
1982b Marginal paradises. Bungalow tourism on the islands of Southern Thailand. *Annals of Tourism Research* 9(2): 189–228.

Cowan, G.
1977 Cultural impact of tourism with particular reference to the Cook Islands. In *A New Kind of Sugar. Tourism in the Pacific*, B.R. Finney and K.A. Watson (eds). Santa Cruz, California: Centre for South Pacific Studies.

Crystal, E.
1978 Tourism in Toraja (Salawesi, Indonesia). In *Hosts and Guests: The Anthropology of Tourism*, V. Smith (ed.). Oxford: Basil Blackwell.

Duffield, B.S., and J. Long
1981 Tourism in the Highlands and Islands of Scotland. Rewards and conflicts. *Annals of Tourism Research* 8(3): 403–31.

Esman, M.R.
1984 Tourism as ethnic preservation: The Cajuns of Louisiana. *Annals of Tourism Research* 11: 451–67.

Farver, J.M.
1984 Tourism and employment in the Gambia. *Annals of Tourism Research* 11(2): 249–65.

Table 3.1 (*Continued*)

Fong, P.
1980 Tourism and urbanization in Nausori. In *Pacific Tourism as Islanders See It*, South Pacific Social Sciences Association and the Institute of Pacific Studies, pp. 87–8. The Institute of Pacific Studies of the University of the South Pacific and the South Pacific Social Sciences Association.

Fukunaga, L.
1977 A new sun in North Kohala: The socio-economic impact of tourism and resort development on a rural community in Hawaii. In *A New Kind of Sugar. Tourism in the Pacific*, B.R. Finney and K.A. Watson (eds). Santa Cruz, California: Centre for South Pacific Studies.

Gamper, J.A.
1981 Tourism in Austria. A case study of the influence of tourism on ethnic relations. *Annals of Tourism Research* 8(3): 432–46.

Greenwood, D.J.
1978 Culture by the pound: an anthropological perspective on tourism as cultural commoditization. In *Hosts and Guests*, V.L. Smith (ed.). Oxford: Blackwell.

Hudman, L.E.
1978 Tourist impacts: the need for regional planning. *Annals of Tourism Research* 5(1): 112–25.

Horoi, S.R.
1980 Tourism and Solomon handicrafts. In *Pacific Tourism as Islanders See It*, South Pacific Social Sciences Association and The Institute of Pacific Studies, pp. 111–14. The Institute of Pacific Studies of the University of the South Pacific and the South Pacific Social Sciences Association.

Jordan, J.W.
1980 The Summer People and the natives. Some effects of tourism in a Vermont vacation village. *Annals of Tourism Research* 7(1): 34–55.

Jud, G.D., and W. Krause
1977 Evaluating tourism in developing areas: an exploratory inquiry. *Journal of Travel Research* 15(2): 1–9.

Kariel, H.G.
1989 Tourism and development: Perplexity or panacea? *Journal of Travel Research* 28(1): 2–6.

Kariel, H.G., and P.E. Kariel
1982 Socio-cultural impacts of tourism: an example from the Austrian alps. *Geografiska Annaler* B64: 1–16.

Kousis, M.
1989 Tourism and the family in a rural Cretan community. *Annals of Tourism Research* 16(3): 318–22.

Lange, F.W.
1980 The impact of tourism on cultural patrimony. A Costa Rican example. *Annals of Tourism Research* 7(1): 56–68.

Loukissas, P.J.
1982 Tourism's regional development impacts. A comparative analysis of the Greek Islands. *Annals of Tourism Research* 9(4): 523–41.

Table 3.1 *(Continued)*

May, R.J.
1977 Tourism and the artefact industry in Papua New Guinea. In *A New Kind of Sugar. Tourism in the Pacific*, B.R. Finney and K.A. Watson (eds). Santa Cruz, California: Centre for South Pacific Studies.

McElroy, J.L., and K. De Albuquerque
1986 The tourism demonstration effect in the Caribbean. *Journal of Travel Research* 25(2): 31–4.

McKean, P.F.
1978 Towards a theoretical analysis of tourism: Economic dualism and cultural involution in Bali. In *Hosts and Guests*, V.L. Smith (ed.). Oxford: Blackwell.

Meleisea, M., and P.S. Meleisea
1980 'The best kept secret': Tourism in Western Samoa. In *Pacific Tourism as Islanders See It*, South Pacific Social Sciences Association and The Institute of Pacific Studies. The Institute of Pacific Studies of the University of the South Pacific and the South Pacific Social Sciences Association.

Monk, J., and C.S. Alexander
1986 Free port fallout. Gender, employment and migration on Margarita Island. *Annals of Tourism Research* 13(3): 393–413.

Niukula, P.
1980 The impact of tourism on a Suvavou village. In *Pacific Tourism as Islanders See It*, South Pacific Social Sciences Association and The Institute of Pacific Studies, pp. 83–5. The Institute of Pacific Studies of the University of the South Pacific and the South Pacific Social Sciences Association.

Peck, J.G., and A.S. Lepie
1978 Tourism and development in three North Carolina coastal towns. In *Hosts and Guests*, V.L. Smith (ed.). Oxford: Blackwell.

Samy, J.
1980 Crumbs from the table? The workers' share in tourism. In *Pacific Tourism as Islanders See It*, South Pacific Social Sciences Association and The Institute of Pacific Studies, pp. 67–83. The Institute of Pacific Studies of the University of the South Pacific and the South Pacific Social Sciences Association.

Skinner, R.J.
1980 The impact of tourism on Niue. In *Pacific Tourism as Islanders See It*, South Pacific Social Sciences Association and The Institute of Pacific Studies, pp. 61–4. The Institute of Pacific Studies of the University of the South Pacific and the South Pacific Social Sciences Association.

Smith, V.
1978b Eskimo tourism: Micro-models and marginal men. In *Hosts and Guests*, V. Smith (ed.), pp. 51–70. Oxford: Blackwell.

Swain, M.B.
1978 Cuna women and ethnic tourism: A way to persist and avenue to change. In *Hosts and Guests*, V. Smith (ed.). Oxford: Blackwell.

Table 3.2 Survey work on tourism community relationships

Ap, J., T. Var, and K. Din
1991 Malaysian perceptions of tourism. *Annals of Tourism Research* 18: 321–3.

Belisle, F.J., and D.R. Hoy
1980 The perceived impact of tourism by residents: A case study in Santa Marta, Colombia. *Annals of Tourism Research* 7(1): 83–101.

Brayley, R., and T. Var
1989 Canadian perceptions of tourism's influence on economic and social conditions. *Annals of Tourism Research* 16(4): 578–82.

Bystrzanowski, J.
1989 *Tourism as a Factor of Change: A Sociocultural Study*. Austria: Vienna Center.

Caneday, L., and J. Zeiger
1991 The social, economic, and environmental costs of tourism to a gaming community as perceived by its residents. *Journal of Travel Research* 30(2): 45–9.

Davis, D., J. Allen, and R.M. Cosenza
1988 Segmenting local residents by their attitudes, interests, and opinions toward tourism. *Journal of Travel Research* 27(2): 2–8.

Dowling, R.K.
1993 Tourism planning, people and the environment in Western Australia. *Journal of Travel Research* 31(4): 52–8.

Haukeland, J.V.
1984 Sociocultural impacts of tourism in Scandinavia. *Tourism Management* 5(3): 207–14.

Johnson, J.D., and D.J. Snepenger
1993 Application of the tourism life cycle concept in the Greater Yellowstone Region. *Society and Natural Resources* 6(2): 127–48.

Johnson, J.D., D.J. Snepenger, and S. Akis
1994 Residents' perceptions of tourism development, *Annals of Tourism Research* 21(3): 629–42.

King, B., A. Pizam, and A. Milman
1993 Social impacts of tourism: Host perceptions. *Annals of Tourism Research* 20(4): 650–5.

Lankford, S.V.
1994 Attitudes and perceptions toward tourism and rural regional development. *Journal of Travel Research* 32(3): 35–43.

Lankford, S.V., and D.R. Howard
1994 Developing a tourism impact attitude scale. *Annals of Tourism Research* 21(1): 121–39.

Liu, J., and T. Var
1986 Resident attitudes toward tourism impacts in Hawaii. *Annals of Tourism Research* 13(1): 193–214.

Liu, J.C., P.J. Sheldon, and T. Var
1987 Resident perception of the environmental impacts of tourism. *Annals of Tourism Research* 14(1): 17–37.

Table 3.2 (Continued)

Long, P., R. Perdue, and L. Allen
1990 Rural resident tourism perceptions and attitudes by community level of tourism. *Journal of Travel Research* 28(3): 3–9.

Madrigal, R.
1993 A tale of tourism in two cities. *Annals of Tourism Research* 20(2): 336–53.
1995 Residents' perceptions and the role of government. *Annals of Tourism Research* 22(1): 86–102.

McCool, S.F., and S.R. Martin
1994 Community attachment and attitudes toward tourism development. *Journal of Travel Research* 32(3): 29–34.

Milman, A., and A. Pizam
1988 Social impacts of tourism on Central Florida. *Annals of Tourism Research* 15(2): 191–204.

Mok, C., B. Slater, and V. Cheung
1991 Residents' attitudes towards tourism in Hong Kong. *International Journal of Hospitality Management* 10: 289–93.

Murphy, P.E.
1981 Community attitudes to tourism: A comparative analysis. *Tourism Management* 2(3): 189–95.

Perdue, R.R., P.T. Long, and L. Allen
1987 Rural resident tourism perceptions and attitudes. *Annals of Tourism Research* 14(3): 420–9.
1990 Resident support for tourism development. *Annals of Tourism Research* 17(4): 586–99.

Pizam, A.
1978 Tourist impacts: The social costs to the destination community as perceived by its residents. *Journal of Travel Research* 16(1): 8–12.

Pizam, A., and J. Pokela
1985 The perceived impacts of casino gambling on a community. *Annals of Tourism Research* 12(2): 147–65.

Prentice, R.
1993 Community-driven tourism planning and residents' preferences. *Tourism Management* 14(3): 218–27.

Ritchie, J.R.B.
1988 Consensus policy formulation in tourism: measuring resident views via survey research. *Tourism Management* 9(3): 199–212.

Schlüter, R., and T. Var
1988 Resident attitudes toward tourism in Argentina. *Annals of Tourism Research* 15(3): 442–5.

Sheldon, P.J., and T. Var
1984 Resident attitudes to tourism in North Wales. *Tourism Management* 5(1): 40–7.

Tsartas, P.
1992 Socioeconomic impacts of tourism on two Greek isles. *Annals of Tourism Research* 19(3): 516–33.

dealing with stage-based approaches or equity based explanations, there is an inconsistency in the evidence.

Overall three major problems exist in this area of research and in combination they provide an explanation for the considerable confusion that appears to exist. First there are definitional and measurement problems with the concepts of tourists, tourism and community. In particular there has been no attempt to investigate systematically the nature of the tourism phenomenon. The available work is mainly targeted to differences in hosts and only a handful of studies has examined differences in tourists or tourism. In the latter case there has been some attention paid to the level of tourism development but little given to the types of tourism. Burr (1991) also observes that the concept of what constitutes a community has not been considered carefully by researchers. In a review of twenty-five empirical studies on tourist impacts and communities Burr attempted to classify the views and approaches to the community concept used by the researchers. He found a good deal of variation in the clarity of the researchers' views of a community. Occasionally researchers appeared to use a simple human ecological model focusing only on the term 'community' as a synonym for place, while a further small set of articles adopted critical elements such as an emphasis on power, decision making or dependency as part of their analysis. It can be argued that a pluralism of theoretical approaches is appropriate to suit a diversity of tourism study purposes. In general, when defining 'community' for tourism community studies researchers should at least contemplate the view that a community is an interacting communicating and dynamic entity.

A second major problem additional to our concern with definitional issues is that of describing and profiling perceived impacts of tourism. While not all of the survey studies have this as an aim most studies ask respondents to rate in some way a list of tourism impacts. Unfortunately very few studies develop these lists from respondents themselves or give their respondents the opportunity to add to or comment on these lists. This issue goes back to the suggestions by Cohen (1979) that good tourism research should be emic in its design, that it should consider the perspectives of the participants, not necessarily the perspectives of the researchers as a part of the research testing process. Since only a few studies have asked respondents to rate or assess the importance of various impacts we also have confusion between the existence of an impact and its scale of importance. Thus it can be argued that we have a very limited view of the nature and content of the hosts' perceptions of tourism both by using an etic or non-emic filter and by not considering the importance of the impact literature. This issue is also related to a limited use of the current psychological research on attitudes and a certain naiveté by tourism researchers in adopting this psychological research for their own studies.

The final and most important problem perhaps is the lack of theory. The atheoretical nature of much tourism research is a major limiting factor in

Table 3.3 Major conclusions of selected survey studies of community perceptions of tourism

Variable examined	*Study*	*Key conclusions*
1. Level of tourism development	1. Liu, Sheldon & Var (1987)	Residents of places with a longer history of tourism development are more aware of both positive and negative impacts.
	2. Allen, Long, Perdue & Keiselbach (1988)	There was a curvilinear relationship between perceptions of negative impacts and development of tourism, but this was not as strong a relationship as between perceptions of negative impacts and population growth.
	3. Long, Perdue & Allen (1990)	There was a curvilinear relationship between support for tourism and level of development, but as levels of tourism development increase so do perceptions of both negative and positive impacts.
	4. Perdue, Long & Allen (1990)	Perceptions of impacts are related to level of tourism development.
	5. Madrigal (1993)	Level of tourism development is the best predictor of perceptions of negative but not positive impacts of tourism.
2. Economic dependency on tourism A. Comparisons of residents, business owners & government officials	1. Pizam (1978)	Entrepreneurs were more positive about tourism than other groups.
	2. Thomason, Crompton & Kamp (1979)	Entrepreneurs were more positive about tourism than other groups.
	3. Keogh (1990)	There were no significant differences in the perceptions of business owners and residents.
	4. Lankford (1994)	Residents were more cautious than business owners and public officials.
	5. Murphy (1983)	There were significant differences between residents, administrators and the business sector.
B. Job in tourism or perceived positive balance of personal costs and benefits of tourism	1. Pizam (1978)	There was a positive relationship between employment in and support for tourism.
	2. Rothman (1978)	Economic dependency on tourism was related to more positive perceptions of tourism.
	3 Husbands (1989)	Residents employed in tourism were more positive about tourism.

	4. Perdue, Long & Allen (1990)	Personal benefits from tourism were important in explaining perceptions of positive but not of negative impacts of tourism.
	5. Mansfeld (1992)	Residents employed in tourism were more positive about tourism.
	6. Madrigal (1993)	Personal benefits from tourism were the best predictors of perceptions of positive impacts.
	7. Prentice (1993)	There was a positive relationship between perceived benefits of tourism and positive perceptions of tourism.
	8. Glasson (1994)	Positive relationship between working in tourism and support for tourism.
	9. Lankford & Howard (1994)	Those who were more dependent on tourism were more positive about tourism.
3. Distance from place of residence to tourist areas	1. Belisle & Hoy (1980)	As distance from place of residence to tourist areas increased residents were less positive about tourism.
	2. Brougham & Butler (1981)	Some relationships were found between residence in zones of high tourist pressure, but the nature of the relationship differed for different types of tourist.
	3. Sheldon & Var (1984)	Residents in higher tourist density areas were more positive about tourism.
	4. Keogh (1990)	People living closer to a proposed tourist development perceived more negative impacts.
	5. Mansfeld (1992)	People living further from tourist areas saw more negative impacts from tourism.
4. Level of contact with tourists	1. Pizam (1978)	Residents with more contact with tourists were negative about tourism.
	2. Rothman (1978)	High contact with tourists was associated with positive perceptions of tourism.
5. Respondent demographics	1. Belisle & Hoy (1980)	No relationships between perceptions of tourism and age, sex or level of education.
	2. Brougham & Butler (1981)	Older residents were less positive about tourism.

Table 3.3 (Continued)

Variable examined	*Study*	*Key conclusions*
	3. Davis, Allen & Cosenza (1988)	No relationships between demographics and attitudes towards tourism.
	4. Ritchie (1988)	Older residents were less positive about tourism.
	5. Husbands (1989)	Education and age were related to perceptions of tourism.
	6. Perdue, Long & Allen (1990)	There were no relationships between perceptions of tourism and demographics when personal benefits from tourism were controlled for.
	7. Caneday & Zeiger (1991)	Level of education was related to more positive perceptions of tourism for residents, but it was related to more negative perceptions for entrepreneurs who were not in tourism.
	8. King, Pizam & Milman (1993)	Only limited differences in perceptions of tourism between different demographic groups.
	9. Lankford (1994)	No significant relationships between perceptions of tourism and demographics.
6. Community attachment	1. Brougham & Butler (1981)	People who had lived longer in a community were more positive about some types of tourists.
	2. Davis, Allen & Cosenza (1988)	People born in a place were more positive about tourism than newcomers to a place.
	3. Lankford & Howard (1994)	No significant relationship between community attachment and perceptions of tourism.
	4. McCool & Martin (1994)	Greater attachment to a community was associated with higher ratings of both positive and negative impacts of tourism.
7. Use of outdoor recreation facilities	1. Perdue, Long & Allen (1987)	No significant differences in perception of tourism between groups with different levels of outdoor recreation.
	2. Keogh (1990)	Residents who used an area proposed for tourism development saw both more positive and negative impacts from the development.

8. General economic conditions of a community	1. Perdue, Long & Allen (1990)	If residents believe the future of their town is bright they are less supportive of tourism development.
	2. Johnson, Snepenger & Akis (1994)	Lower support for tourism was related to low levels of general economic activity.
9. Perceived ability to influence tourism decisions	1. Madrigal (1993)	Perceived ability to influence decisions was significantly related to positive perceptions of tourism.
	2. Lankford & Howard (1994)	There was a significant positive relationship between perceived ability to influence tourism decisions and perception of positive and negative impacts of tourism.
10. Knowledge of tourism	1. Davis, Allen & Cosenza (1988)	Knowledge of tourism was positively related to positive perceptions of tourism.
	2. Keogh (1990)	Greater knowledge of a proposed tourism development was associated with more detailed and more positive perceptions of tourism impacts.
	3. Lankford & Howard (1994)	Greater knowledge of tourism was related to greater support for tourism.
11. Political self-identification	1. Snepenger & Johnson (1991)	Residents with conservative political views were more negative about tourism than those with moderate or liberal views.
12. Influence of a tourism public relations campaign	1. Robertson & Crotts (1992)	Residents of a community with a public relations campaign were more positive about tourism than those not exposed to a public relations campaign.

Note: A number of survey studies were not included in this table because they did not examine the relationships between any variables and perceptions of tourism. Rather these studies provide descriptions of community responses. For example, Pizam and Telisman-Kosuta (1989), Getz (1994), Cooke (1982), Andressen and Murphy (1986), Murphy (1981), Pizam and Pokela (1985) and Milman and Pizam (1988).

its development. Some of the tourism community researchers recognize this issue and Ap (1990) goes so far as to say that 'unless researchers launch out of the elementary descriptive stage of current research and into an explanatory stage where research is developed within some theoretical framework they may find themselves none the wiser in another ten years time' (p. 165).

Clearly this review of existing work on tourism and communities (and it is limited to the work published in English) provides several imperatives for directing further attention to this topic. Kurt Lewin, an eminent social scientist, was fond of saying 'there is nothing so practical as a good theory'. Lewin's comments could be paraphrased for this topic area as 'there is nothing so needed as a good theory'. Such a theory would need to consider carefully what kinds of tourists, tourism and communities are being studied, it would need to have the power to explain and integrate some of the existing research findings and to have substantial credibility in the broader realm of social science.

The challenge then for the tourism community relationship research is clear. We need to reformulate this area of study; we need to ask new questions or develop a new theoretical perspective so that we do not pursue an endless list of unconnected studies using different definitions which fail to provide a cumulative body of knowledge.

It is time therefore to return to our analysis of tourism questions. Let us consider again the issue of depth. Are we asking our questions at sufficient depth? Are we really exploring the way communities function in relation to the tourism issue or are we simply working in what might be identified as tenure research mode, where we pick relatively accessible variables and ask simple questions about them? Additionally, we need to ask whether our questions are well focused? Are we really clear as to why we are conducting these studies? Are planners and communities benefiting from this work or is it a self-serving academic literature? One particular insight from this attention to relevance is that we must communicate our findings about community reactions in a way that planners and tourism managers can understand; it is community views and meaningful community groups which we need to identify, not a parade of isolated variables. As we return to our agenda for good questions the possibility that we can look for a new theoretical approach and shift the kind of question we are asking would seem to be an appealing opportunity.

The theoretical approach which will be used to add insight to this area is that of social representations theory. This approach explicitly takes an emic perspective and seeks to understand the reality of the social actor. Social representations are clusters of attitudes about a socially significant topic but they are more than that. Moscovici (1981, 1984) refers to social representations as everyday theories or branches of knowledge for organizing the world. Social representations may be characterized by a set of defining phrases and images such as 'Aids is a gay plague' or 'Tourism is a vulture destroying cultures'. These organizing phrases and images are a crystallization of linked social and political attitudes as well as a strong perceptual filter influencing the way individuals and groups see the world. Key features of social representations theory are represented in Table 3.4.

A social representations emphasis for the tourism community relationships area has some exciting logical possibilities. Social representations have been used in other social science fields to understand how whole com-

Table 3.4 Key features of social representations theory

1. Social representations are complex meta systems of everyday knowledge and include values, beliefs, attitudes and explanations.
2. The content and structure of social representations are important.
3. Social representations help to define and organize reality.
4. Social representations allow for communication and interaction.
5. Social representations make the unfamiliar familiar.
6. Through the use of metaphors, analogies and comparisons with prototypes social representations fit new and abstract concepts/events into existing frameworks.
7. Images are central components of social representations.
8. Abstract concepts are both simplified (through the use of images and analogies) and elaborated (through connections to existing knowledge).
9. Social representations have an independent existence once created and so can be found in social or cultural artefacts.
10. Social representations are critical components of group and individual identity.
11. Social representations are important features of group interaction and so social representations theory explicitly recognizes social conflict and the importance of power in social dynamics.
12. Social representations are prescriptive. They can direct both action and thought (especially perception).
13. Social representations are not deterministic or static. They vary along many dimensions including the level of consensus about them, their level of detail and how they are communicated. Individuals can and do influence, create and change social representations. They can be changed through individual influence, direct experience, persuasive communication and/or group interaction.
14. Social representations connect individuals to their social/cultural worlds.
15. Social representations are both influenced by and influence science.

munities feel about such topics as health, sex and madness. Tourism would seem to be an equally appealing topic for the attention of a social representations framework. It has already been suggested that social representations are a system of knowledge shared by a community. This view challenges all the existing logic of community tourism studies. The existing logic is that the impacts of tourism cause people to have an attitude towards tourism. Social representations suggests that the way we view tourism, that is our system of knowledge or social representation of it, affects the way we perceive and respond to impacts. This suggests a redirecting of our questions. Perhaps we have to start asking how people view tourism overall and then ask the question how does this overall view of tourism affect impacts rather than perceiving the causal chain to be in the other direction. Additionally, a further redirecting of our questions comes from close observation of the evidence. There is, contrary to some reviews, no simple consistent summary of the individual sociodemographic variables in the survey studies. Sometimes working in tourism makes people very pro tourism; sometimes they are less enthusiastic about new developments. Our observation of the data suggests other forces are at work in terms of how people view and define tourism. The question of relevance is also germane to this discussion. The needs of communities, of governments, of industry and planners suggest that we should redirect and sharpen our questions. People come in whole packages, not slices of ages, genders, occupations and residential locations. What is desirable for the relevance question and for the planners is to know who shares what views and, by analogy, what community segments (akin to market segments) exist for tourism development attitudes. With these directions we can turn our attention to asking questions not of difference about community attitudes nor even questions of description, the two kinds of questions which characterize the earlier phases of tourism community relationship study, but questions of pattern. Thus, in a more extensive discussion of this topic, existing data on Hawaii and New Zealand have been reviewed and it has been demonstrated that these large social representations, that is macro views of tourism, do exist (Pearce, Moscardo and Ross 1996). Additionally, we have collected original data and analysed that material for a number of northern Australian cities and again been able to show that the large-scale views of tourism are driving the views of impacts rather than the reverse.

What is a typical social representation of tourism? In the Hawaiian example one of the strong, consistent views shared by some community members is that tourism is an *engine of growth, an economic powerhouse* for the whole economy. For the New Zealand sample there was some initial evidence that when tourism was seen as an industry which developed recreational opportunities for people 'like New Zealanders' then it was positively received. The social representation here is of tourism as *town supporter and saviour*. The Australian studies also identified a group of people who thought that tourism was a force which neglected them – a view that tourism was an *enemy of the disadvantaged* – and there were those who were *culture concerned* in terms of tourism being an entity which changed the way local life could be lived. There is a myriad of smaller patterns in the attitudes and perceptions underlying these broad descriptions of the way tourism is conceptualized. It is important to appreciate that the social representations perspective has strong links to people's social identity and that group identification concerns may shape which representation people adopt for their system of knowledge about tourism.

Armed with this framework it is no longer possible to see the attitudes of tourism as isolated, atomistic and reductionist points in the conceptual space of the resident. Rather, tourism attitudes are a part of a larger community representation of the way industries and important social phenomena are perceived. These broad social attitudes are fuelled by the media as well as by everyday conversation. The logic and content of social representations involves communication, discussion and social interaction.

In terms of our question-asking analysis we have moved over time in the tourism community research field from questions of description to questions of difference. It can now be suggested that by exploring questions of pattern we may be able to generate a richer research field for the future.

The specific concern of this chapter, better question asking in the research field of tourism community relationships, can be extended to a wider circle of influence. In particular the social representations framework, which was the theoretical underpinning of the present chapter and is highly relevant to questions such as what do people share in their views on a specific topic, could be extended to areas such as tourist destination images, as well as key social phenomenon for tourists such as health, crime, sex and climate. More generally, the value of tackling tourism research agendas with a rich question-asking framework is likely to lead researchers away from an impoverished single-discipline approach to a topic and towards a programme of studies focusing on different levels of question, different and similar participant views of a topic and questions varying in focus and having long term and short-term relevance. As research on tourism development expands we could paraphrase Lord Florey's queries to his students – it is not just a matter of what question you are trying to answer but what suite of questions, for whom and in what time frame.

Acknowledgment

The authors acknowledge the Cooperative Research Centre for Tropical Rainforest Ecology and Management as a partial funding body for this research.

References

Ap, J. (1990) 'Residents' perceptions research on the social impacts of tourism', *Annals of Tourism Research* 17 (4): 610–16.

Bickel, L. (1983) *Florey: The Man Who Made Penicillin*, Melbourne: Sun Books.

Burr, S.W. (1991) 'Review and evaluation of the theoretical approaches to community as employed in travel and tourism impact research on rural community organisation and change', pp. 540–53 in A.J. Veal, P. Jonson, and G. Cushman, *Leisure and Tourism: Social and Environmental Changes*, papers from the World Leisure and Recreation Association Congress, Sydney, Australia, 1991.

Butler, R.W. (1980) 'The concept of a tourist area cycle of evolution; implications for the management of resources', *Canadian Geographer* 24 (1): 5–12.

Cohen, E. (1979) 'Rethinking the sociology of tourism', *Annals of Tourism Research*, 6: 18–35.

Doxey, G.V. (1975) 'A causation theory of visitor resident irritants: methodology and research inferences', pp. 195–8 in *The Impact of Tourism, Sixth Annual Conference Proceedings*, Travel Research Association, San Diego.

Faulkner, B., Pearce, P., Shaw, R., and Weiler, B. (1995) 'Tourism research in Australia: confronting the challenges of the 1990s and beyond', pp. 3–25 in *Bureau of Tourism Research, Tourism Research and Education in Australia, Proceedings from the Tourism and Educators Conference, Gold Coast (1994)*, Canberra: Bureau of Tourism Research.

Hawkins, D.E. (1993) 'Global assessment of tourism policy', pp. 175–200 in D.G. Pearce and R.W. Butler (eds) *Tourism Research. Critiques and Challenges*, London: Routledge.

Krippendorf, J. (1987) *The Holiday Makers: Understanding the Impact of Leisure and Travel*, London: William Heinemann.

Moscovici, S. (1981) 'On social representations', pp. 181–209 in J.P. Forgas (ed.) *Social Cognition: Perspectives on Everyday Understanding*, London: Academic Press.

—— (1984) 'The phenomenon of social representations', pp. 3–70 in R.M. Farr and S. Moscovici (eds) *Social Representations*, Cambridge: Cambridge University Press.

Murphy, P.E. (1981) 'Tourism course proposal for a social science curriculum', *Annals of Tourism Research*, VIII (I), 96–105.

—— (1985) *Tourism: A Community Approach*, New York: Methuen.

—— (1988) 'Community driven tourism planning', *Tourism Management* 9 (2): 96–104.

Pearce, P.L. (1995) 'From culture shock and culture arrogance to culture exchange: ideas towards sustainable socio-cultural tourism', *Journal of Sustainable Tourism* 3 (3): 143–54.

Pearce, P.L., Moscardo, G.M., and Ross, G.F. (1996) *Tourism Community Relationships*, Oxford: Elsevier.

Ritchie, J.R. (1993) 'Tourism research: policy and managerial priorities for the 1990s and beyond' pp. 201–16 in D.G. Pearce and R.W. Butler (eds) *Tourism Research. Critiques and Challenges*, London: Routledge.

Ritchie, J.R. Brent and Goeldner, C.R. (eds) (1994) *Travel, Tourism, and Hospitality Research: A Handbook for Managers and Researchers*, New York: John Wiley & Sons, Inc.

Ryan, C. (1995) *Researching Tourist Satisfaction: Issues, Concepts, Problems*, New York: Routledge.

Schonland, A.M and Williams, P.W. (1996) 'Using the Internet for travel and tourism survey research: experiences from the Net Traveler Survey', *Journal of Travel Research*, 25 (2): 81–7.

Walle, A.H. (1996) 'Tourism and the Internet: opportunities for direct marketing', *Journal of Travel Research*, 25 (1): 72–7.

WTO (1997a) *World Tourism Leaders' Meeting: The Social Impacts of Tourism. Final Report*, Madrid, Spain: World Tourism Organization.

—— (1997b) *Tourism 2000: Building a Sustainable Future for Asia–Pacific. Final Report.* Asia–Pacific Ministers' Conference on Tourism and Environment, Maldives, 16–17 February, Madrid, Spain: World Tourism Organization.

4

TOURISM ACADEMICS AND TOURISM PRACTITIONERS

Bridging the great divide

Carson L. Jenkins

INTRODUCTION

The purpose of this chapter is to discuss the proposition that academics and academic publications seldom, if ever, influence tourism development and how it is implemented. A first objective is to invite comment and debate arising from the proposition. A second objective is to consider ways in which the divide between academics and practitioners might be bridged.

The proposition has arisen from the field experience of the author who, although a full-time academic, has experience of tourism policy, planning and development extending for over twenty-five years in approximately forty developing countries. The proposition emerges therefore from participant observation in tourism development, supplemented by an analysis of the tourism development literature. It may be said at the outset that the proposition may be construed as an outburst of empirical whimsy rather that an objective analysis supported by an extensive literature review which is the basis of most academic writing. In defence of the participative observation approach, it can be argued that there is no direct writing on this subject, although some published literature (WTO 1994) indirectly supports the contention.

It is also necessary to dispel any assumption that academics do not engage in consultancy work; many do so but, as is argued here, their involvement tends to be as *specialist technicians* rather than as formulators of the planning approach. It is further argued that the planning approach to tourism development is usually derived from experience, is methodologically standardized and is little influenced by academic publications. There are tourism planning exercises and projects where this assertion may not be applicable: however, as a generalization the contention is that it is valid.

The author's empirical experience is rooted in the developing countries where most tourism development which includes planning, policy, and implementation strategies is funded by international development agencies and carried out by professional consulting companies. There is a fundamental difference between consultants' implementation-oriented work and academic research. One of the main differences in output is that 'client confidentiality' usually limits the publication and distribution of consultancy-based reports, whereas academic research is primarily aimed at publication to support the dissemination of knowledge. These issues will be further discussed below.

THE GREAT DIVIDE

The two groups which are the focus of this chapter are tourism academics and tourism practitioners.

Tourism academics

Although tourism as an activity has a long tradition (Lickorish and Jenkins 1997), as an academic subject it is a development of the 1960s. Initial concentration of the subject development was in Europe but it is now seen as a worldwide phenomenon. In the 1960s there were few academic institutions which offered tourism courses (as differentiated from travel trade and hospitality specific programmes). Consequently, there were only a few academics who could term themselves as 'tourism lecturers'. However, as tourism expanded as both a domestic and international activity, it inevitably attracted the interests of academics from many disciplines with a consequent development of courses and degree programmes. Tourism as a subject is now found in many academic institutions throughout the world and many academics now designate themselves as 'tourism lecturers'. It is now possible for an academic to foresee a career in the academic study of tourism with a range of institutions in many countries offering a varied career path. The growth of tourism as a subject is reflected in the continuously expanding number of publications related to the subject in books, specialist journals, conferences and increasingly through postings on the Internet.

As a group, academics have three main characteristics based on their traditional role of being thinkers, researchers, analysts and teachers:

- to advance both the knowledge and understanding of the subject;
- to disseminate information via teaching, publications and conferences;
- and, through teaching and publications, to educate and influence students, other academics and the industry.

Although there may be some debate about these characteristics they do seem to represent what most academics do. In the area of teaching the purpose is not only to inform students but also to influence their thinking by equipping them with concepts and techniques which are relevant to their future careers, an objective supported by the growing literature on tourism.

The need to publish in academic journals and through subject-specific books seeks to influence (if not always impress!) the peer group – fellow academics. The dissemination of information is essentially career-motivated and scholarly. Some of the central and continuing debates, for example defining the subject area (Jafari 1990; Leiper 1981; Tribe 1997), the nature and definition of 'sustainable' tourism (Butler 1991; Cater 1993; Forsythe 1996; Hunter 1997), the role of tourism in society (Smith 1989; Wilkinson 1987; Wilson 1997) are all topics which advance our theoretical understanding of the subject but are unlikely to be of much interest to practitioners. A further difficulty is that given the need for academic rigour in publications and the review system, many publications become dated and lose their usefulness before they are published. There may be a further problem where tourism practitioners do not read the academic press. It is suggested, from the basis of experience, that this is often the case. It follows therefore that academics write for and seek to inform fellow academics. Much of the innovative work being done by academics will tend to have a very limited influence on tourism development practice unless the academic is also a practitioner.

For example, the extensive and growing literature on ecotourism, on community participation in tourism development planning, and host–guest relationships are three areas of current interest to academics. Although these issues are also of concern to practitioners their concerns are essentially pragmatic, resource-constrained and implementation-oriented. Of necessity, practitioners must be focused within the parameters of the project and do not seek more general prescriptions.

Practitioners

In relation to development planning, tourism practitioners are either individual consultants or consulting companies. Consultants, unlike academics, operate in a business environment and may be identified as having three broad operational characteristics.

- Their work is contractual, project specific and profit driven.
- Knowledge dissemination is through project-specific reports, plans and studies which are commissioned by the client and have a limited circulation.
- They aim, by the development of expertise and reputation, to secure future contract work.

Consultants by the very competitive nature of their work do not readily make available to enquirers their past reports unless these are likely to be influential in securing future work. As most projects are in two stages, that is project preparation and subsequent implementation, consultants tend to protect their client base and are always, like any business, seeking to ensure their company sustainability through further contracts. Much tourism development planning is funded by international agencies, for example World Bank, European Union, United Nations Development Programme, and, for regional development, by banks such as the Asian Development Bank. Many projects are large, long-term and require a variety of technical skills inputs. Few academic institutions would have either the resources or flexibility to compete for large contracts with the major consultancy companies in the area of tourism development.

For these reasons, where large contracts are available many consulting companies subcontract specialist requirements to individuals, often academics and sometimes academic institutions. In this way, there is a 'cross-over' between the academic and the practitioner.

Cross-over

Although it is postulated that academic institutions cannot realistically or effectively compete in the international tourism contract market, academics are employed as *project specialists* offering skills and experience usually on a short-term basis, for example tourism economists, forecasters and marketing specialists. However, these posts are filled after the contract has been signed and the project strategy and methodology agreed. A brief description of the project preparation stage will support this point.

Project preparation

Although projects vary in scale and complexity there are basically *five* steps in the project cycle which are briefly discussed here (EU(a) 1993):

1 Project identification

   Depending on the nature and scale of the project there are usually three sub-components involved in this stage:

   (a) Technical specification
   At this stage of project identification the concept, for example to develop a tourism resort, or to specify a tourism planning exercise, is formalized into a project proposal. The project's objectives, activities and expected outputs (results) are identified. The aim is to establish wherever possible measurable indicators to eventually evaluate the project if it should go ahead.

   (b) Feasibility studies
   For some projects a feasibility study will be done to assess the viability of the proposal. For international lending agencies a

particular area of interest will be to determine how the proposed project fits into any indicative programme of general guidelines for co-operation between the beneficiary country and the lending institution. For example, what priority is given to the tourism sector in a particular country (perhaps indicated in a national development plan) and to what extent will the proposed project contribute to national and sectoral development aims? In some African countries, such as Zambia, tourism is not only recognized as a priority sector by the Government, but as Zambian tourism is almost exclusively based on wildlife viewing, which obviously takes place in rural areas, tourism development contributes to increasing rural development, incomes and employment – all national economic objectives (EU(b) 1995).

(c) Financial proposal

If the feasibility study confirms the viability of the proposed project and the overall objectives meet national development objectives, then a financial proposal will be drawn up for the project. At its simplest, the financial proposal sets out the budget for the project.

At the project identification stage it is usual to employ a consultant(s) to prepare the various working documents. The technical specification and the financial proposal are used to prepare the Terms of Reference.

2 Terms of Reference

The Terms of Reference is the project 'bible'. It sets out in detail what the project aims to achieve, what activities are necessary, what outputs (results) are expected within a given budget. Usually incorporated into a comprehensive Project Document, the Terms of Reference not only guides the consultants undertaking the project but is also the main evaluating instrument to check performance against specification.

3 The tender process

Consultants may be invited to bid for a contract either on an open or restricted basis. The open system means that, with perhaps some qualification, any company can bid for the contract. A restricted tender operates where only selected companies are invited to bid. In the latter case this may be as a consequence of time pressures or more likely due to the specialist nature of the contract.

The bids received will usually have three dimensions: a profile of the tendering company (including a record of previous experience); a technical proposal to indicate how the contract specifications are to be met (i.e. project methodology and the proposed team of consultants); and a financial proposal. It is not always the case that the lowest bid price wins the contract. Evaluations of tenders should involve careful scrutiny of past experience, an evaluation of the proposed methodology and of the team of consultants assembled to carry out the project, and of the financial proposal.

The evaluation of tenders is a fairly mysterious process. The quality of the evaluation will depend very much on those doing it – their knowledge, experience and their understanding of the proposed project. Some project evaluations are guided by attempts to systematize the process often by using a points-scale reference to assess different project components. In some cases the process may proceed along more subjective lines. The outcome of the tender process is to award and sign a contract with a consultant(s).

4 Project implementation
The consultant implements the project by using the financial resources provided in the contract. Implementation will require the consultant to prepare a plan of operation and to present reports on the progress of the project as specified in the contract. It is also usual to appoint a project Steering Committee to oversee the operation of the contract and to monitor and discuss any issues arising. The Steering Committee is the main operational channel between the consultant/contractor and the client.

5 Project evaluation
This involves the review and analysis of results and the impact of the project. Project evaluations can take place during or after a project has ended. In the case of a multi-stage project, the start of the next phase will usually depend on the conclusions of the previous stage.

Although the outlined project schematic may vary between agencies depending on a variety of circumstances and practices, the main stages are common to development planning exercises. What is important to note is that the methodology for the project is devised by the bidding company/consultant.

METHODOLOGY

Despite the increasing funds which are being channelled to tourism planning and development, very little is known about how project methodology is chosen, and on what basis. For example in the WTO report (1994) a brief synopsis of various country development plans is given for country projects, but there is no discussion of how a particular methodology was chosen, what alternatives might have been considered, and how this selection was made. Although much has been written on micro-methodologies, such as cost–benefit analysis, economic impact studies, social impact analysis, environmental assessment techniques, little has been written on macro-project methodology for tourism. Methodological issues are usually conceptual, for example the debates on sustainability, on community participation and empowerment through decentralization. It is these conceptual but essentially pragmatic issues which are often project-relevant and are the main focus for academic writing and research. For many academics the issue is to ask 'why?' rather than 'how'; the latter question could be interpreted to be the main concern of the consultant.

It is suggested that in most projects, methodology is derived from past experience and it is the tried and tested methodology which prevails over anything innovative. In projects where profit margins are often tight, innovation might be costly. For example if, as is now generally agreed, community participation in tourism development is desirable, this might be written into the Terms of Reference. How it is to be implemented is a challenge despite the extensive academic writing on this topic (Murphy 1985; Pearce 1994; Ritchie 1993). In practice, participation may only involve consultation, and perhaps the consultant employed to undertake this task (academic or otherwise) may be methodologically constrained by the project manager, or the budget, or both.

In most large projects there are a series of methodological issues. At the project (or macro) level the main concern is to devise a methodological approach to cover the activities and to produce the outputs required within both the time-scale and budget. At the component (or micro) level, for example economic impact study, marketing survey, analysis of demand and forecasting, the same constraints apply but methodology and techniques are more firmly established. At all times, methodology is driven by pragmatism and is informed by experience rather than theory.

Although methodology is an area where one would expect academics to be most comfortable, it appears that few academics actually are involved in this at the project level. Some of the important contributions to tourism in the developing countries remain enshrined in the literature often unread by consultants. This statement is of course impressionistic and is based on the author's experience. It is a contention which might be tested via a survey of tourism practitioners or from responses to this chapter. A more important source of information could be derived from the financiers of tourism development and in particular, the institutional lenders.

Although organizations such as the European Union, the United Nations Development Programme, development banks and country-based donor organizations have and continue to fund tourism projects, very little information has been released about the relative success or failure of these projects. These lenders have the ability, and probably need, to create *institutional memory*, that is, to compare through project appraisal and evaluation a major databank of tourism-related information. Such information is not only relevant to the prospects of future projects but, if disseminated, would become a stimulant to debate. Such an innovation would have to come from a project's lender; client confidentiality would preclude the consultant from revealing such information.

Methodology has a particular significance in tourism because of its multidisciplinary nature. Various aspects of a single project can include economic, social, environmental and political inputs, each of which might require a specific methodological approach which will have to be harmonized within the operational framework of the project to meet the Terms of Reference. Some of the major current debates in tourism development include: sustainability; the environment; social impacts; community participation; local empowerment; type, scale and location of development; globalization; indigenization; women in development and host–guest relationships. These are issues which have attracted major academic interest and publications – how can tourism practitioners and funding agencies be made more aware of this output? How can it be fed into project methodologies?

From the brief discussion of the project cycle it can be deduced that there are two points where the methodological approach can be influenced. First is at the stage of preparing the Terms of Reference for the project and second, at the implementation stage where the project's activities are undertaken. As noted previously, the Terms of Reference set out the activities, inputs and expected outputs from the project. To a considerable extent the writer of the Terms of Reference can exert a direct influence on methodology by specifying certain activities which are integral to the project, for example that the planning process must incorporate some degree of community involvement; that the scale of the project take into account possible impact on the social and natural environment; that measures be devised to increase participation of women in the development process. In each case the Terms of Reference will not normally specify how these activities will be done – but that they have to be incorporated into the implementation or operational phase. Well-written Terms of Reference can incorporate within a project most of the concerns that are of current academic debate – for example, how is the sustainability of the project to be achieved?

The Terms of Reference are also informed by the client's perspective and any legal and planning framework perspective which might exist in the project's locale. Where clients are international development agencies they will insist on certain development parameters which directly affect project methodology. For example, many development agencies, such as the World Bank, require that more development initiatives have to be channelled to and through the private sector. Long experience of state-dominated capitalism, for example in both Eastern European and African countries, has not generated the expected levels of development benefits. In African countries, for instance Zambia, a 'privatization unit' is selling off the state's commercial assets such as copper mines to international and local private investors.

Where part of a project's remit is to support more private sector involvement in the development effort specific measures have to be devised to do this. In both Zambia and Zimbabwe, although private sector companies were and are involved in the tourist sector, there was no representative body which had the recognized stature to provide an 'industry view'. In both countries, aided by development assistance, industry representative associations were initiated in the Tourism Council of Zimbabwe and the Zambia Tourism Council which in the latter case is being formally incorporated within the proposed Tourism and Hospitality Bill.

In terms of community participation in tourism development the Zimbabwean 'Campfire' project provided benefits for village communities in return for anti-poaching and wildlife conservation activities. This type of initiative meets the requirements for more local participation in tourism development and also helps to increase rural development particularly related to employment and income. From these examples it can be seen how the Terms of Reference of a project can ensure that general development objectives are included within the parameters of a specific project.

The Terms of Reference guide the formulation of project methodology – how the project is to be implemented. It is useful here to make a distinction between project methodology and project management. The methodology determines 'how to' and project management monitors, changes and ensures that time-scales are adhered to. As noted above, very little is known about how project methodology in tourism is formulated. It is determined at the contract tender stage where the methodological proposal itself becomes an important aspect of the tender. In fairly large projects, such as national development plans, experience suggests that methodology is derived from the 'tried and tested' approaches. These types of plans tend to break down into well-recognized components: marketing, transport, human resources development, environmental impact etc. To conform to the activities of the project and to meet the expected outcomes as specified in the Terms of Reference, little innovation is required. The main opportunities for methodological innovation are related to the component activities. It is in these areas of activities where academics can become practitioners themselves through subcontracted inputs to the project or through academic writing which informs practitioners of developments in the field. This is the 'great divide' and where little is known about 'cross-over' activities. To what extent are academics involved in tourism development projects as advisers or as temporary practitioners? How can they influence project methodology and/or implementation? Are tourism practitioners aware of the latest issues/developments/concepts in the field? It is these questions, largely unanswered, which should be incorporated into a new agenda.

BRIDGING THE GAP – ISSUES FOR A NEW AGENDA

If it is accepted that there is a gap between tourism academics and tourism practitioners then how might it be bridged to the mutual benefit of both parties? The following suggestions are not exhaustive but hopefully focus on some of the fundamental concerns raised by this chapter.

Research

At a country level it should not be difficult to identify those consulting companies with interests in and experience of tourism development projects. By means of a survey and probably personal interviews the issue of the formulation of methodology could be explored, as could the relevance and knowledge of academic publications on tourism. Information might also be gathered on the links with and use of academics in projects. A carefully structured research effort could yield information to lighten what is at present a black hole.

Publications

As noted above, the review system often has the effect of considerably delaying publications. In an era where the electronic media predominate, it should become increasingly possible to 'post' at least preliminary data and draft articles on specific networks accessible both to consultants and academics. There are the problems of copyright and plagiarism. However, it is one means of getting work more widely and quickly disseminated.

Another problem is that many academics would say this approach is essentially one-sided; what would the consulting companies offer in return? It may be that this type of information distribution would provide the consultancy companies with more information of what academic interests are, and who is working in what field – which may also lead to consulting opportunities for more academics. This would facilitate 'cross-over' but some academics might argue that this would tend to compromise their objectivity as commentators on and analysts of the tourism sector. The required outputs of the project might limit if not preclude debate and experimentation on the more theoretical and conceptual issues which academics are mostly interested in.

Meetings

It is noticeable that in most academic conferences on tourism issues – of which there is a growing number – very few consultants or consulting companies attend. This may be due to pressure of work or perhaps because the practitioners do not have much interest in what academics are discussing? More effort should be given to inviting and involving practitioners in these fora, probably by asking for a written or/oral presentation. Creation of more common interest will again facilitate 'cross-over'. For many younger academics it may be an occasion to understand what the consultancy imperatives are.

Education

In an increasingly professional era, most consultants will have received a university-level education. The proliferation of tourism programmes and the growth of the industry worldwide, is creating more awareness of the career potential in tourism. Academics have an opportunity to influence students with a balanced view of tourism, how it has developed and what issues might have importance to its future. A well-rounded education in the subject area should become the basis for life-long learning. The knowledge base must be continually refreshed. Academics can help to do this. The task would be facilitated if more of the practitioners' output could 'cross-over'. Unfortunately, as most of the practitioners' work is client-confidential, it will require dispositions from the clients to release non-sensitive project materials. For example, the large number of national tourism development plans which have been completed rarely find their way into the public domain. These project outputs could become very useful teaching aids and perhaps shed light on some of the methodological issues raised in this article.

Academic participation

There are four areas where academics can influence tourism projects. First, is where they are invited to write the Terms of Reference for a project. Second is where they are employed by a consulting company to help devise the methodology for a project tender. Third is at the implementation stage usually involved as a specialist (e.g. economist, environmentalist). Fourth, where the academic is appointed to the project's Steering Committee which monitors the progress of the project. It is often possible through this involvement to input to the project methodology or to suggest changes as necessary and appropriate.

Participation will require that the academic has a reputation in the specialist field. Herein lies a problem. Although the academic may be perceived by his or her peers as a leader in the field, the practitioner may yet regard him as a citizen of an 'ivory tower'. Grossly unfair though this designation may be, tourism academics who do not engage in practical project work may find their influence marginalized or not even heeded at all. In part of course, this situation does derive from the nature of the academic's work which often involves theoretical and conceptual thinking which may not have an immediate application.

CONCLUSION

There does seem to be empirical evidence that there is a divide between tourism academics and tourism practitioners. There is a postulated need to facilitate more 'cross-over' and some measures have been suggested in this article. However what is important is to recognize that there is a synergy between the tourism academic and the tourism practitioner. There is a possibility of increasing mutual benefit, an objective which both groups should strive to achieve.

Essentially, tourism development is all about *implementation*. Sound or even innovative projects have no impact on development unless they are implemented. Perhaps too much of the concern of academics is about *impact* rather than the *process* of implementation. The two aspects are inextricably linked because the process of implementation is itself a methodological issue which can directly affect the impact of what is done. To have a more central role in formulating methodology would allow the academic a real-world participatory function which should benefit the project, and the experience gained would also benefit teaching and future research. There are considerable 'cross-over' benefits to be gained by both parties if the great divide can be effectively bridged.

References

Butler, R. (1991) 'Tourism, environment and sustainable development', *Environmental Conservation* 18: 201–9.

Cater, E. (1993) 'Ecotourism in the Third World: problems for sustainable development', *Tourism Management* 14: 85–90.

European Union (a) (1993) *Manual of Project Cycle Management: Integrated Approach and Logical Framework*, Brussels: European Union.

—— (b) (1995) *Zambia, Medium-Term National Tourism Strategy*, Lusaka: EU.

Forsythe, T. (1996) *Sustainable Tourism: Moving from Theory to Practice*, Godalming: World Wildlife Fund/Tourism Concern.

Hunter, C. (1997) 'Sustainable tourism as an adaptive paradigm', *Annals of Tourism Research* 24 (4): 850–67.

Jafari, J. (1990) 'Research and scholarship: the basis of tourism education, *Journal of Tourism Studies* 1 (1): 33–41.

Leiper, N. (1981) 'Towards a cohesive curriculum in tourism: the case for a distinct discipline', *Annals of Tourism Research* 8 (1): 69–84.

Lickorish, L.J. and Jenkins, C.L. (1997) *An Introduction to Tourism*, Oxford: Butterworth–Heinemann.

Murphy, P. (1985) *Tourism: A Community Approach*, New York: Methuen.

Pearce, P.L. (1994) 'Tourism: resident impacts; examples, explanations and emerging solutions', in W.F. Theobald (ed.) *Global Tourism*, Oxford: Butterworth–Heinemann.

Ritchie, J.B.R. (1993) 'Crafting a destination vision; putting the concept of resident-responsive tourism into practice', *Tourism Management* 14: 5.

Smith, V.L. (ed.) (1989) *Hosts and Guests: The Anthropology of Tourism*, 2nd edn, Philadelphia: University of Pennsylvania.

Tribe, J. (1997) 'The indiscipline of tourism', *Annals of Tourism Research* 24 (3): 638–57.

Wilkinson, P.F. (1987) 'Tourism in small island: a fragile dependence', *Leisure Studies*, 6: 127–46.

Wilson, D. (1997) 'Paradoxes of tourism in Goa', *Annals of Tourism Research* 24 (1): 52–75.

World Tourism Organisation (1994) *National and Regional Tourism Planning: Methodologies and Case Studies*, London and New York: Routledge.

5

PROBLEMS AND ISSUES OF INTEGRATING TOURISM DEVELOPMENT

Richard W. Butler

INTRODUCTION

Integration is a term which has a number of meanings in the context of tourism. While it is common now to call for 'integrated tourism planning' (Haywood 1988; Innskeep 1991), there has been a well-established pattern of integration with respect to the supply side of tourism since tourism became a popular activity. Tourism in its modern form is perhaps a century and a half old, if the emergence of mass tourism can be taken as being represented by the first use of mass transportation in an organized fashion for leisure travel. The early co-ordination of rail and coach travel, commercial accommodation and guide services by Thomas Cook, the father of modern tourism, represents the first formal integration of services in tourism (Swinglehurst 1982). While individual tourists made use of a range of tourist services which developed from the period of the Grand Tour onwards, few of these were integrated in a formal sense and most developed in an *ad hoc* manner over the course of time in response to regular demand by a small but affluent group of travellers (Towner 1985). It is from the early days of mass tourism that we can see the emergence of the idea of integrating facilities and services for tourism. However, such integration was purely on the operational or supply side of tourism and was predominantly the work and responsibility of tour operators such as Cook and those who followed his example. For much of the next century the integration which existed in tourism was related to transportation and accommodation in particular, or, putting it another way, to getting the tourist to and from the destination and accommodating them when they were there.

The development of the familiar and common array of facilities found at many of the mass tourist destinations of the late nineteenth century, resorts such as Blackpool, Brighton, Atlantic City, Ocean City, Manley, Nice, Cannes or Santander, which have given rise to their distinctive resort morphology or Recreation Business District (Stansfield and Rickert 1970) was a later occurrence. The apparent integration of amenities and facilities found at these destinations also tended to come about in an incremental rather than a planned and organized fashion, with a few exceptions such as the Boardwalk at Atlantic City. Only in the classic spa towns of Europe such as Bad Gastein or Baden Baden, and to some degree in their North American equivalents such as Saratoga Springs, was there evidence of planning in an integrative manner. Bath, in south-west England, is perhaps the ultimate example of integrated planning and management of tourism in an urban setting, under the almost tyrannical control of Beau Nash, with facilities not only being developed but also operated in a highly efficient and socially rigorous manner. In the case of Bath and Nash's guidance, the emphasis was clearly upon facilities serving the social requirements of the elite audience rather than meeting appropriate resource and environmental needs of the destination, as is more the case in the present day. In some forms of modern tourism destination development we have perhaps almost come full circle from the stylized integrated form of development of Bath and the elegance of Nash to the enthusiasm of the GOs of the Club Med, and the formalized entertainment offerings of the Pump Rooms being replaced by water sports and competitions. In the last two decades in particular much tourism development, especially in developing countries, has been characterized by the integrated resort. Such developments represent tourist enclaves in destination areas, and are designed on the principle of supplying such a range of attractions and services that the tourists staying there need not leave the development. All facilities, operations and attractions are integrated on the one site (Innskeep 1991). Holiday camps in the United Kingdom represented one of the first forms of integrated resorts when they first appeared in the 1950s, and the Club Med type of development operates on a similar principle. In more recent years the scale of development of integrated resorts has become much larger and the nature of many resorts much more luxurious and up-market. While having certain advantages in terms of potential environmental and quality control operations, such developments have been criticized for minimizing tourist expenditures in local communities and high levels of economic leakage. The fact that many of the larger developments are externally owned and often import much of their food and other items aggravates this situation. This form of integrated tourism development, however, is not the focus of this chapter.

The discussion here is concerned with the concept of integration on a larger scale, namely that involving the integration of tourism itself, and its elements, with its surroundings in destination areas, rather than integration within tourism in the manner discussed above. It will be argued in the chapter although there is a great deal of positive discussion about integration of tourism and in particular integrated planning of tourism, that in fact true integration of tourism is not common, and that this lack of appropriate integration represents a major reason for the often considerable dissatisfaction with tourism which emerges over time in many destinations. The chapter proceeds by discussing first the basic principles of integration, followed by the policy issues which it creates. A review of the difficulty of integrating tourism precedes a commentary on integration in practice. The chapter concludes with a discussion of the problems inherent in achieving successful integration.

PRINCIPLES OF INTEGRATION

Integration has become a common and frequently used term in the tourism literature, particularly in the context of planning and development. Apart from Innskeep's (1991) volume relating specifically to the topic, the term finds considerable reference in writing by authors such as Gunn (1988), Haywood (1988), Murphy (1985), Pearce (1989, 1995), and Wall (1997) on general issues relating to tourism planning and development. In the more specific area of sustainable tourism, for example, we can find the term being supported in a considerable number of publications such as those by Bramwell *et al.* (1996), Priestley, Edwards and Coccossis (1996), Nelson, Butler and Wall (1993), Stabler (1997) and Wahab and Pigram (1997). This is not surprising, as the concept of sustainable development implies the linking or integration of elements of the environment in an appropriate manner, Wahab and Pigram (1997: 4) commenting that 'sustainability is an integrative concept'.

Although the term is widely used, like sustainable development, it is rarely adequately defined. A dictionary definition describes the verb to integrate as meaning 'to make or be made into a whole; incorporate or be incorporated' (Collins 1988: 588), and goes on 'to amalgamate or mix . . . with an existing community'. In the context of tourism, therefore, we can regard integrated planning and development as meaning the process of introducing tourism into an area in a manner in which it mixes with existing elements. It is implicit in such an understanding that this introduction and mixing is done in an appropriate and harmonious way, such that the end result is an acceptable and functionally successful community, in both ecological and human terms. Successful integration in many cases is similar to the basic principles so eloquently presented by Ian McHarg (1967) in his book *Design with Nature* over three decades ago. As pressure from supporters of Agenda 21 to incorporate tourism development with the principles of sustainable development increase, the successful integration of tourism with other processes and activities in destination areas becomes more significant and important.

It is possible to note at least three principles which should make the concept of integration in the context of tourism appropriate and appealing to those involved in tourism development:

- acceptability
- efficiency and
- harmony

The successful integration of tourism development into a community or destination should make the development more acceptable to local residents and existing resource users than when a tourism development is imposed in a segregated and unrelated manner. Successful integration can also be a much more efficient process. To most planners, developers and managers efficiency is a goal to be achieved, and steps and processes which speed the process of completion and acceptability of development are to be welcomed. As important, perhaps, is the fact that successful integration can avoid problems that could otherwise materialize later in the operational phase of development through achieving synergy and even symbiosis (Budowski 1976) with other existing activities and resource processes rather than competition or conflict (Butler 1993). Finally, at a time when conflict resolution is an increasingly common process in development, the achievement of harmony and the avoidance of conflict through appropriate and acceptable development is another situation that is desired by planners and developers alike.

Despite this general support for the concept of integration in tourism development and planning, particularly in the last decade, in reality it is hard to find true examples on the ground. In many cases there are significant gaps between the concept as described in many plans and what actually appears and is operationalized in specific tourist destinations. There are a number of reasons which can be put forward for this pattern of events which are discussed below. It can be argued that it is often not a lack of intent nor a lack of desire to achieve integration, but rather like the implementation of sustainable development, more a case of a lack of understanding and information about the nature of tourism which prevents the goal being achieved.

POLICY ISSUES

As with any process relating to development, the successful achievement of goals depends on a variety of factors. In the context of integration and its inclusion in planning there are a number of what may be termed policy issues which have to be taken into account. The integration of tourism development implies inevitably that tourism is being added to a set of one or more existing activities and processes in the particular destination. Of key importance are the intra-relationships within this set of activities and processes and potential inter-relationships between them and tourism.

- *Priorities*. In the real world very few economic or social-cultural activities are equal in priority. There is normally a constant state of competition or at best uneasy alliance between activities in most communities, some being in the ascendance and others stagnating or being in decline. In many parts of the world in which tourism has been introduced, traditional primary activities have had a long established priority, both in terms of economic importance and because of their often intricate links with cultural patterns and behaviour. Tourism is often seen as being in direct competition with these traditional activities and in such situations integration rather than imposition is essential if development is to be successful.
- *Control*. In many areas in which tourism is being introduced or expanded control of resources and governance is often vested in those community members involved in other activities. In such situations tourism may be seen as a threat to the established power structure, as a force which seeks to obtain control of resources and space. Thus acceptance may be difficult if tourism is not carefully and appropriately fitted into the existing systems. This situation is often compounded by the fact that in the case of large-scale tourism developments, ownership and control of these developments commonly lies outside the local community, and the element of external control can make acceptance of such development much more problematic (Sofield 1996). Lack of appreciation of local preferences, patterns of activities and priorities can exacerbate such a situation (Berno 1996).
- *Scale*. Clearly, small-scale developments can generally be integrated into communities and systems more easily than large complex developments which make major demands on local resources. Tourism development occurs in a wide range of scales from generally small local operations such as the provision of accommodation in private homes (Wall and Long 1995) or itinerant selling of souvenirs (Cukier 1996) to large, often foreign-owned inclusive resorts with major space and labour requirements.
- *Stage or timing of development*. Unless tourism is being introduced into an area devoid of settlement, the destination will contain a variety of activities at various stages of development and can itself be categorized as being developed, developing or undeveloped (Pearce 1989). Even where tourism development is taking place in an unsettled area, consideration has to be taken of natural processes. In a community the stage of development can be of crucial importance with respect to the probability of the successful integration of tourism. Mature, complex or

sophisticated (in the economic sense) communities have a greater chance of successfully integrating tourism development into their pattern of activity and absorbing the economic impacts of tourism than more basic economies, where the effects may be much more severe and unanticipated.

- *Community harmony or conflict.* Communities rarely have uniform views, particularly on factors such as tourism that are seen as capable of causing considerable change to and within the community. Tourism, as other forms of development, is capable of generating the full range of emotions from enthusiastic support to outright antagonism, not only over time as in Doxey's (1976) over-simplistic unidirectional model, but also from the initial suggestion of tourism before development ever occurs (Mathieson and Wall 1982). The issue may be not so much how feelings towards tourism develop over time, but whether there are opposing views at any time. There need to be policies and practices in communities to determine views towards development and to enable the community to decide what type and scale of development can be accommodated in the community before development occurs (Sofield and Birtles 1996). If such processes are in place and operational, then integration of the appropriate type of development should be much more easily achieved.

Other issues are also involved in determining whether tourism development will be successfully implemented in a particular community or not, and many are place and time specific, as well as relating to the nature of the specific development.

PRACTICALITY

While the issues discussed above are perhaps more abstract and general in nature, there is also a range of practical issues which significantly affect the likely success or failure of the integration of tourism in specific locations.

- *Multiplicity of form.* As Simpson and Wall note in Chapter 14, tourism comes in many forms and its effects are extremely varied. What may be appropriate procedures for the integration of tourism development of one form may not be suitable or successful for another form. Furthermore, it is rare for a destination to experience only one form of tourism or one form of tourism development during its life cycle. Thus the successful integration of one type of tourism does not guarantee that all development will be accepted and integrated in the same way, or that new forms of tourism will be able to be incorporated into the community as successfully. Different sectors, even within the same form of tourism, can place very different demands upon the destination resource base with varying levels of acceptability.

- *Dynamics*. In many respects the dynamic nature of tourism is the most difficult and greatest problem to be overcome in achieving the successful integration of tourism. Rates of development and change are themselves inconsistent, and the faster the rate of development the more difficult the task of integration because the time available for consultation with the community and the study of ecological processes in the destination is reduced. As has been discussed elsewhere (Butler 1997) the agents of change themselves are subject to change, causing an infinite variety of inconsistency. Thus the integration process has to be dynamic and flexible itself, and capable of adapting to changing circumstances. The dynamism in the system extends to views about tourism, as well as change in economic activities and processes, in cultural behaviour, and in political viewpoints (Hall 1994).
- *Benefits and costs*. It is not unreasonable to assume that if the attributes noted above accrue from the successful integration of tourism, then they can be regarded as benefits and will have a tangible economic form. An absence of conflict, increased speed and efficiency of planning and development and then of operation, coupled with an absence of post-development adjustment because of inappropriate scale of development, all have potential social and environmental benefits as well. However, the fact that benefits can accrue from integration does not necessarily by itself ensure the success or viability of the development. In many cases the benefits may be derived mostly or even exclusively by the developers and external agencies, allowing for dissatisfaction and opposition to continue or to develop at the local level. Benefits generally have a mirror effect, in that benefits in one area or at one level may be matched by costs in other areas and sectors. Successful integration needs to ensure that costs at the local level are at least matched by the benefits at the same level, and hopefully that they accrue to the same sectors which face costs because of the development. In many cases the benefits of tourism may accrue to new actors on the economic scene, while to other established operators there are only costs incurred by accommodating tourism development or expansion. If the principles of sustainable development (WCED 1987) are to be implemented in tourism development, then equity will become a major consideration, applying to the allocation of benefits and costs in the present as well as in the future.

In addition to the above factors, other elements such as politics and the origins of the actors involved can become significant influences on the success or failure of attempts to integrate tourism into a community. In reality the political structure and relationships within a community and with external agencies and levels of government can create strange associations and groupings of views (Hall 1994). This can be complicated further by the origins of proponents and opponents of development and the ways in which tourism could be integrated into a community. The varying backgrounds, familiarity with the destination, and different levels of knowledge about tourism and the context for development can all affect the success or failure of attempts at integration.

PRACTICE

The gap between the successful integration of tourism and partial or symbolic integration can be wide. A great deal depends on the attitudes of those with responsibility for tourism planning and development and the approaches which they take to co-ordinate and include the full range of viewpoints on proposed developments including those of local residents. It has been argued that if local participation is to be effective in overcoming the often inherent problems of entrenched decisionmaking and inequality in the allocation of benefits from development, then it must originate within the local communities themselves (Mowforth and Munt 1998). The level of involvement and the ability to bring about changes in development proposals so that the final results are truly integrated into the local systems vary widely. One of the first researchers to describe the various stages of participation was Arnstein (1969), who proposed six 'steps' in her 'Ladder of Public Participation'. More recently, Pretty (1995) has described seven stages in a typology of participation. The five categories noted below reflect the same basic elements of level of participation and decision-making power vested in the communities involved, as do Arnstein's and Pretty's typologies. They range from 'imposition' through to 'equality', reflecting to a large degree the nature of decision making in the host country and community.

- *Imposition*. Clearly the least desirable, and probably the least effective, form of integration is where it is forced upon a community without input from that community. In such cases the development is normally controlled and financed from outside the community and has the support of external and higher levels of government. The Nusa Dua development on Bali serves as an example of such a type of development. A complex of international franchise hotels and associated facilities developed by external financing with strong support from the central Indonesian government was constructed over a period of years in southern Bali in the 1980s (Wall 1993). The resort can be said to be integrated on the basis that it has strong economic links through transportation developments and employment with other parts of Bali, is promoted as a part of Bali, not as a separate destination, and to some extent has integrated water and other elements of environmental planning and operations. It is spatially segregated from the rest of Bali by virtue of its location and design, and by the market at which it aims, which is considerably higher than that attracted by most of the rest of Bali.

- *Petition.* Whereas the previous category implied a total absence of local involvement in the decisions on location, scale, design and other elements of integration of development, this category suggests an opportunity for input. However, that input is made without obligation on the part of the decision-making agencies to accept such input or to modify their intentions. In these cases the proposal is presented to local communities for comment only after demands and petitioning for such an opportunity has been made and often at a point at which major decisions have been made. The opportunity for integration is mostly one-sided, that is, local operators and resource users may have an opportunity to adjust their patterns of operation to allow integration with the development before it is begun, or at least completed, thus allowing for an easier and possibly more successful insertion of the development. To some extent this situation reflects the process at Anuha Island discussed by Sofield (1996), where the situation deteriorated rapidly following changes in ownership and operation and the perceptions of the level of involvement of local decision makers were abruptly changed with disastrous results.
- *Advice.* This category represents a situation in which local advice is sought as to the ways in which the development could be designed, located and operated in order to make it more integrated into the community, although the final decision-making authority still remains outside the community. There is the opportunity for local expertise and priorities to be reflected in the development, not only in its scale and operation, but also in its design and location. Integration that is effective is more likely to be achieved at this level and beyond in that there is for the first time the opportunity for adjustment to achieve integration on both the part of the developers and of the local communities. This process was followed by National Park authorities in Canada in the 1970s when public participation was introduced to park planning for the first time.
- *Representation.* At this stage the communities have formal rights to be involved in the process of integration through planning and operation of the development. This may take the form of representation on local boards or agencies to oversee the design, construction and even operation of the development. With involvement from an early stage local expertise can be incorporated more fully into design and operational procedures and the effects of development and use on the environment

be minimized. The greater the sharing of knowledge about the local environment and the operation of the development the greater the opportunity for avoiding conflict and incompatibility as tourism develops. In Canada, tourism developments in those national parks which have permanent settlements are now designed and operated in such a way, and the residents are actively involved from the early stages (Parks Canada 1994). Final decisions are still taken outside the local area, and the process can still cause concern and generate opposition, because compromises still have to be made.

- *Equality*. True equality in the integration of tourism developments is rare. It can be taken to represent the situation in which the proponents and recipients of development are treated as equals and all elements of the integration process are agreed to by all parties. Adjustments in the design and operation of developments are as likely, or perhaps even more likely than adjustments in the operation of existing activities and lifestyles. Decision making rests in both locations, the communities and the source of the development. Clearly, where development is coming from within the community and the community operates on a democratic or equally acceptable format for its residents, then this situation should be automatic. Where development is proposed from outside, then procedures need to be in place to ensure that proponents are required to treat the communities as equals. A workable veto is obviously one way of achieving this situation. It is likely that such a situation or something equivalent may materialize in Nunavut, the soon to be established Inuit controlled section of the Canadian North West Territories, where the local governing body will not only have control of development but also the funds to carry it out (Milne, Grekin and Woodley 1998).

The categories which have been discussed above represent generalized levels of local involvement in decision making, and have been discussed in a variety of forms with respect to planning, but have not been applied specifically to the integration of tourism developments. As will be discussed below, one of the major problems is that it has taken a long time for the idea of local involvement in the tourism planning process to become a reality in many areas (Murphy 1985). In many parts of the world, unfortunately, a large proportion of which are the scenes of major tourism development and expansion, it still does not exist. However, the involvement of local actors in planning does not necessarily carry over into involvement in development and even less so in operations. Integration is much more than planning, however, and involves all elements from design, planning, construction and operation if it is to be truly successful. The presence of local involvement does not guarantee successful integration, but one may argue strongly that its absence is very likely to result in *unsuccessful* integration.

PROBLEMS

If it is accepted that current thinking and attitudes with respect to tourism development are strongly in favour of an integrative approach, and that such an approach is very much in line with the principles of sustainable development, which also appears to have widespread political and popular support, it is perhaps surprising that examples of the successful integration of tourism are hard to find. Pearce's (1995) example of such an approach in Sarawak is one case where external advisers have gone much further than most in clearly expressing the need for and utility of integration. The Integrated Rural Tourism Development Programmes for Andalusia, discussed by Cavaco (1995) represent another. In reality, while the intent may be present in other areas, although one may question whether there is little more than symbolic intent in some situations, there are a number of factors which make implementation difficult to achieve.

- *Lack of equality*. Despite long-standing calls for local and community involvement in tourism (Murphy 1985), echoed strongly in *Our Common Future* (WCED 1987), local communities do not have the same power as higher levels of government or large-scale external development agencies. Local activities, especially where these are small-scale and highly fragmented, for example small family farms or individually owned fishing boats, can rarely be effectively and strongly represented when large-scale developments are being promoted and discussed. Different priorities are involved at different levels of government and local concerns may be subordinated to national or regional level policies. Such a situation can often mean that while tourism may be integrated in planning at national and regional levels, in reality it becomes imposed at the local level. Efforts may be made to integrate it at the community level, but in such situations existing activities are more likely to have to be adjusted to accommodate tourism than tourism significantly modified to fit in with established local activities. Such situations are common to development of all types. It is rare for local interests, generally represented by a small and limited power base, to be able to successfully achieve and retain equality with higher levels of government and large external private sector corporations. When they are able to do so, it is often because specific local conditions such as geographical attributes and timing, along with the contributions of key individuals, provide the means to achieve such equality of power (Butler and Nelson 1994).
- *Lack of desire*. To achieve successful integration can require considerable resources of time and money, as well as consistent commitment. In most countries elected representatives have relatively short-term viewpoints, often only as far as the next election. To ensure proper integration requires involvement from the first stage of the initial proposal for

development through to monitoring the operational phase of developments. Even if the same decision-makers and officials were in power throughout that period, their priorities and commitments are likely to change and their concerns for specific developments in specific areas likely to wane after development has occurred. There are marked similarities with the issue–attention cycle of impact analysis. The environmental impact analysis process often attracts considerable attention in the early stages of a project, but decreasing attention is generally paid as the project continues, and in very few cases does monitoring continue and rarely are the original projections of impact reviewed and verified once the development is in place and operating. Resources are rarely allocated to ensure that integration of development continues into the operational phase, even if integration was achieved during the planning and development of a project. Development rarely stops when initial construction is completed and the dynamic element of tourism noted above almost always results in modifications to the physical and operational characteristics of most projects over time. Such modifications should also be reviewed and assessed as to their likelihood of being successfully integrated into the local environment.

- *Lack of appreciation*. Despite all of the research and attention given to tourism over the past two or three decades, there is still considerable ignorance of its effects and the way in which it operates at all levels. The blindly optimistic view of tourism as an economic panacea promoted in the 1960s (Zinder 1969), when it was seen as the salvation of declining areas and thus pushed forward at almost any cost, has fortunately changed as the costs as well as the benefits of tourism are being more accurately assessed. However, to some extent the pendulum has swung too far the other way, and tourism is now often envisioned as a supplementary form of economic activity which can be easily accommodated in many peripheral or marginal areas with little adjustment. It is often assumed that tourism will automatically be compatible with existing activities, for example, that local farmers will supply hotels with food and local fisherman have ready markets for their catches, ignoring the fact that many hotel chains prefer to buy in bulk from offshore suppliers to ensure reliability and consistency in supply and quality, and that many tourists demand foodstuffs that cannot be produced locally. Tourism is a complex, resource-specific industry operating at a global scale, often with scant regard for local priorities. Failure to appreciate that it is an industry with global links, dependencies and relationships can result not only in an inability to achieve integration, but ultimately the replacement of a weak monoculture by one that is strong and dominant, at least in the short term.
- *Lack of mechanisms*. There is a general lack of appropriate mechanisms for ensuring the integration of tourism. There is widespread

acknowledgment of the fragmentation and complexity of tourism, and the difficulty of exerting consistent control over the industry, along with its general failure to successfully regulate itself. For many other economic activities there are established agencies which are designed and empowered to control such elements as production, quality, price, marketing and participation, for example, agricultural marketing boards such as the Wheat Board in Canada, the White Fish Authority in Great Britain or Atomic Energy Authorities in many countries. No such bodies exist for tourism, and control and regulation is left with agencies not normally specifically equipped for the task. Planning authorities in most countries at the national level are either concerned with broad strategies or with promotion of development or both, and control and regulation including integration are found at regional and local levels if they exist at all. Whether tourism is sufficiently unique to require a specific mechanism to ensure its regulation and integration is uncertain. What is certain is that without appropriate mechanisms (along with the desire to enforce them, as noted above) tourism is not likely to be effectively integrated on principles of equity at even the planning stage, let alone the operational stages of development.

- *Lack of data and knowledge.* The lack of detailed knowledge of the way that tourism relates to and impacts on other activities has been noted above. One of the main reasons for this lack of knowledge relates to the lack of data on tourism and on the environment in which it operates. Traditionally data have been collected and analysed sectorally, which has mitigated against a holistic and integrated approach. In addition, data were often from one limited time period only, making it extremely difficult to identify and attribute changes over time to any particular causal agent. Monitoring change was not only expensive and time consuming, but often impossible on the scale necessary. Such change as was identified was often limited to small areas and was rarely placed in a regional, let alone a national context. Much data were collected for administrative units rather than for natural ecosystems, which made identification of causation and scale of effect much more difficult also. Only in recent years has it become possible, through the use of remotely collected data, for example from satellite imaging systems, to collect data at a scale suitable to be able to understand and monitor change in natural systems. Even then, analysis of such data was difficult until suitable techniques such as Geographic Information Systems became available and were applied to tourism (Butler, Boyd and Butler 1999). The advent of a wide range of relatively user-friendly systems for the spatial analysis of data combined with increasing ease of access to large scale data means that one of the major blocks to effective integration planning may be removed.

CONCLUSIONS

The concept of integration is a popular one in the tourism literature and calls for its adoption have increased significantly with the popularization and support for the concept of sustainable development. Successful sustainable development inevitably requires the application of an integrative approach, since by definition it demands that the development process not eliminate future options nor prevent the continued operations, especially of indigenous peoples. In discussing planning the management of tourism for the future, Maitland (1998: 2) comments:

> There is increasing interaction between tourism and other industries and between visitors and host populations unconnected with tourism. The consequence is that we need tourism planning which integrates tourism development into broader strategies for the development of places, and of non-tourism businesses.

As tourism becomes tied ever more firmly to global processes and agents, so it will increasingly be involved with many activities and processes unrelated directly to leisure, and it will only be successful if its development is integrated with these other processes. While failure of a tourism enterprise at the global level through a lack of integration may be unfortunate, tourism development at a local level which is not suitably integrated with local activities and processes can be disastrous (Butler and Hall 1998). The complete integration of tourism development is unlikely in many communities, just as the attainment of sustainability is more of a goal than a reality in most situations. However, sincere attempts at integration which include the involvement of local communities are more likely to be well received than development for which no effort is made to reach compatibility, if not symbiosis, with local ecological, economic and social systems. Even where sustainability is not an issue, development which is a part of a community is generally more successful than development which is apart from a community.

References

Arnstein, S.R. (1969) 'A ladder of citizen participation', *American Institute of Planners Journal* 35: 216–24.

Berno, T. (1996) 'Cross-cultural research methods: content or context? A Cook Islands example', pp. 376–95 in R.W. Butler and T. Hinch (eds) *Tourism and Indigenous Peoples*, London: Routledge.

Bramwell, B., Henry, I., Jackson, G., Prat, A.G., Richards, G., and Straaten, J. (1996) (eds) *Sustainable Tourism Management: Principles and Practice*, Tilburg: Tilburg University Press.

Budowski, G. (1976) 'Tourism and environmental conservation: conflict, co-existence or symbiosis?', *Environmental Conservation* 3(1): 27–31.

Butler, R.W. (1993) 'Integrating tourism and resource management: problems of complementarity', pp. 221–36 in M. Johnson and W. Haider (eds) *Communities, Resources and Tourism in the North*, Thunder Bay, Canada: Centre for Northern Studies, Lakehead University.

—— (1997) 'Modelling tourism development: evolution, growth and decline', pp. 109–28 in S. Wahab and J.J. Pigram (eds) *Tourism Development and Growth: the Challenge of Sustainability*, London: Routledge.

Butler, R.W., Boyd, S.W., and Butler, R.D. (1999) 'The application of geographic information systems in tourism planning and management', *Tourism Management* (forthcoming).

Butler R.W. and Hall, C.M. (1998) 'Conclusion: the sustainability of tourism and recreation in rural areas', pp. 249–58 in Butler, R.W., Hall, C.M. and Jenkins, J. (eds) *Tourism and Recreation in Rural Areas*, Chichester: Wiley.

Butler, R.W. and Nelson, J.G. (1994) 'Evaluating environmental planning and management: the case of the Shetland Islands', *Geoforum* 25 (1): 57–72.

Cavaco, C. (1995) 'Rural tourism: the creation of new tourist spaces', pp. 127–90 in A. Montanari and A.M. Williams (eds) *European Tourism*, Chichester: Wiley.

Collins (1988) *The Collins Concise Dictionary*, London: Collins.

Cukier, J. (1996) 'Tourism employment in Bali: trends and implications', pp. 49–75 in R.W. Butler and T. Hinch (eds) *Tourism and Indigenous Peoples*, London: Routledge.

Doxey, G. (1976) 'A causation theory of visitor–resident irritants: methodology and research inference', pp. 195–8 in *The Impacts of Tourism*, Proceedings of 1976 The Travel Research Association Conference, Salt Lake City: The Travel Research Association.

Gunn, C. (1988) *Tourism Planning*, New York: Taylor and Francis.

Hall, C.M. (1994) *Tourism and Politics: Policy, Power and Place*, Harlow: Longman.

Haywood, K.M. (1988) 'Responsible and responsive planning in the community', *Tourism Management* 9 (2): 105–18.

Innskeep, E. (1991) *Tourism Planning: An Integrated and Sustainable Development Approach*, New York: Van Nostrand Reinhold.

McHarg, I. (1967) *Design with Nature*, New York: Doubleday.

Maitland, R. (1998) *Planning the Management of Tourism for the Future*, Paper presented at Symposium on Planning the Management of Tourism for the Future, University of Westminster, London, March 1998.

Mathieson, A. and Wall, G. (1982) *Tourism: Economic, Social and Physical Impacts*, London: Longman.

Milne, S., Grekin, J., and Woodley, S. (1998) 'Tourism and the construction of place in Canada's eastern Arctic', pp. 101–20 in G. Ringer (ed.) *Destinations: Cultural Landscapes of Tourism*, London: Routledge.

Mowforth, M. and Munt, I. (1998) *Tourism and Sustainability: New Tourism in the Third World*, London: Routledge.

Murphy, P.E. (1985) *Tourism A Community Approach*, London: Methuen.

Nelson, J.G., Butler, R.W., and Wall, G. (1993) *Tourism and Sustainable Development: Monitoring, Planning, Managing*, Waterloo, Canada: University of Waterloo.

Parks Canada (1994) *National Parks Policy*, Queen's Printer: Ottawa.

Pearce, D.G. (1989) *Tourist Development*, 2nd edn, London: Longman.

—— (1995) 'Planning for tourism in the 1990s: an integrated, dynamic, multiscale approach', pp. 229–44 in R.W. Butler and D.G. Pearce (eds) *Change in Tourism: People, Places, Processes*, London: Routledge.

Pretty, J. (1995) 'The many interpretations of participation', *In Focus*, 16: 4–5.

Priestley, G.K., Edwards, J.A., and Coccossis, H. (1996) *Sustainable Tourism? European Experiences*, Wallingford: CAB International.

Sofield, T.H.B. (1996) 'Anuha Island Resort, Solomon Islands: a case study of failure', pp. 176–202 in R.W. Butler and T. Hinch (eds) *Tourism and Indigenous Peoples*, London: Routledge.

Sofield, T.H.B. and Birtles R.A. (1996) 'Indigenous Peoples' Cultural Opportunity Spectrum for Tourism (IPCOST)', pp. 369–434 in R.W. Butler and T. Hinch (eds) *Tourism and Indigenous Peoples*, London: Routledge.

Stabler, M. (1997) (ed.) *Tourism and Sustainability: Principles to Practice*, Watlingford: CAB International.

Stansfield, C.A. and Rickert, J.E. (1970) 'The Recreational Business District', *Journal of Leisure Research* 2(4): 213–25.

Swinglehurst, E. (1982) *Cook's Tours: The Story of Popular Travel*, Poole: Blandford Press.

Towner, J. (1985) 'The Grand Tour: a key phase in the history of tourism', *Annals of Tourism Research* 12 (2): 297–335.

Wahab, S. and Pigram, J.J. (1997) (eds) *Tourism Development and Growth: the Challenge of Sustainability*, London: Routledge.

Wall, G. (1993) 'International collaboration in the search for sustainable tourism in Bali, Indonesia' *Journal of Sustainable Tourism* 1 (1): 38–47.

—— (1997) 'Sustainable tourism–unsustainable development,' pp. 33–49 in S. Wahab and J.J. Pigram (eds) *Tourism Development and Growth: the Challenge of Sustainability*, London: Routledge.

Wall, G. and Long, V. (1995) 'Small-scale tourism development in Bali', pp. 237–58 in M.V. Conlin and T. Baum (eds) *Island Tourism: Management Principles and Practice*, Chichester: Wiley.

World Commission on Environment and Development (WCED) (1987) *Our Common Future*, Oxford: Oxford University Press.

Zinder, H. (1969) *The Future of Tourism in the Caribbean*, Washington DC: Zinder and Associates.

6

ANALYSING HERITAGE RESOURCES FOR URBAN TOURISM IN EUROPEAN CITIES

Myriam Jansen-Verbeke and Els Lievois

'THE URBAN QUESTION'

The development and management of urban tourism has become a major challenge in many European cities. They are cradles of history and culture, of social and cultural life, rich in heritage buildings, but in many cases also areas in which economic, social and environmental problems are concentrated (Cortie, Dekker, Dignum 1993). This 'urban question' now has a high priority on the political agenda, at the level of the European Union (EU) and at the level of countries, regions and cities (EC, DG Regional Policy 1992). The development of urban tourism is expected to rebalance the urban economy, by generating new activities and by regenerating rundown districts. However this 'economic injection' is not by definition, and surely not in all cities, a magic cure. The revitalization potential of tourism strongly depends on the presence of resources which can be developed into tourist attractions, on the financial capacity of public and private partners to do so and on political will. There are significant differences between the European cities in the way urban tourism policies are conceived and integrated in the overall urban development strategies (EC, DG Regional Policy and Cohesion 1997).

Policies for urban revitalization are strongly inspired by the possibilities of exploiting the cultural potential of urban historic districts. In the first stage of the European 'Urban Pilot Project' (1989–93) the main focus was on the regeneration of historic centres (EC, DG XVI 1997). The pilot projects were a first step in the search for solutions to counter the problems of decaying urban centres, which were assumed to have an intrinsic historic value. External support seems to be highly necessary in situations where it is not possible to adapt the historic centres to new uses and space claims, to maintain the heritage buildings in a new functional setting. The high cost of maintaining the physical fabric, the lack of local investments and know-how to break through this circle of decline are also key issues. Renewing and redeveloping run-down commercial areas and improving the quality of the buildings are the new priorities on the political agenda of many historic cities in Europe.

This explains the growing interest in heritage resources, cultural identity (local and regional), the investments in cultural infrastructure and in particular, the search for new uses for old forms. Gradually the emphasis is being laid on the need for an integrated approach to urban regeneration and the role of tourism. This characterizes the second phase of the 'Urban Pilot Programme' of the EU, approved in July 1997 (EC, DG XVI 1997). The EU programme supports innovation in urban regeneration and planning and introduces the challenge of promoting economic and social cohesion at different scales. The key issues of urban revitalization through tourism are summarized in Figure 6.1. The call for new initiatives was a real incentive for many cities in Europe to reconsider their development strategies, including the development of urban tourism and the conservation and reuse of heritage buildings. In particular, local authorities in small and medium-sized cities woke up and are now 'discovering' new opportunities in their place. The area-based approach to the development of tourism is widely introduced and seen to be the best way towards integrated planning.

In this context of political interest in urban revitalization through tourism, amongst other uses, and in view of recent EU incentives, there is a growing need for a critical assessment of the tourism potential of historic cities. The key issues in this process of developing tourism as a vehicle for urban revitalization are the way heritage resources are being transformed into tourist products, the concern to conserve the cultural identity of places and, not least, the major challenge to develop views and strategies on urban quality management. The search for integrated management of quality in urban tourist destinations has caught the attention of public authorities and

1. Transformation of heritage resources
    into urban tourist products
2. Cultural identity of European cities
    Sense of place
    Selection process and interpretation
3. Quality management
    Integration of tourism in the urban system
    Resource management
    Visitor management

Figure 6.1 Key issues in urban revitalization through tourism.

researchers (Jansen-Verbeke and Van de Wiel 1995). The focus of policy supporting research is now on the identification of quality criteria from different points of view, on examples of success stories – where urban tourism management has indeed led to an overall improvement of the destination quality.

URBAN TOURISM

Taking into account the many challenges related to the development of urban tourism in European cities, which go far beyond the marketing of urban tourism destinations, Directorate General XXIII of the European Union decided to constitute a working group on 'Urban Tourism in Europe' in 1997 to support the current debate on the 'urban question' and to define the key issues. The European Commission is funding several actions in favour of European tourism. These affect in various ways current development plans in tourism and the focus of research. One of the objectives is to raise the quality of tourism in Europe by promoting sustainable tourism (Hinch 1996). This implies support to local initiatives concerning visitor management programmes.

Current issues in urban tourism development

In a highly urbanized Europe, cities play an important role in economic competition and in the process of social and cultural cohesion. The high priority of urban revitalization has reopened the debate on the potential and actual role of urban tourism as a 'new' economic activity (Law 1992). Key considerations concerning the development of urban tourism include:

1. The issue of increasing leisure mobility related to a change in time-use patterns (day tourism, short breaks, weekend tourism) and space use patterns in urban and countryside destinations (Cazes and Potier 1996).
2. A reaffirmation of urban attractions, strongly supported by a renewed appreciation of heritage settings, cultural events and a wide range of cultural facilities and activities, including diverse opportunities for shopping and amusement (Law 1993; Herbert 1995).
3. The current process of 'touristification' is irreversible and affects the urban system in physical, economic, social and cultural terms. New infrastructure (e.g. expansion of the hotel sector), and new functions (leisure, amusement, entertainment, shopping) take the place of traditional urban functions. The risks of upsetting the balance of the urban system by a dominance of the tourism function are real (Jansen-Verbeke 1997). Multifunctionality and the profitability of urban enterprises become a major concern in this new trend towards 'Leisure Cities'.

4 The leisure and tourism function tends to claim increasingly more public urban space, and to have a significant impact on the urban environment. This eventually affects the environmental quality for other uses and users.

The market for urban tourism in Europe

Urban tourism in Europe is a growth market, benefiting in particular from the development of short stays and the splitting of holidays. 'It has been established that contrary to other forms of highly seasonal tourism, urban tourism is relatively stable over time – principally as a result of the complementarity between leisure tourism and business and conference tourism' (EC, DG XXIII 1998). The relative importance of urban tourism is reflected in the Eurobarometer survey of the holiday patterns of Europeans (European Commission 1998). Urban destinations account for about 17 per cent of all the holidays of Europeans, which is still well below the coast and mountains, although there are significant differences from country to country (Table 6.1).

Table 6.1 Type of tourist destination of European holidaymakers in per cent of holidays by country. N=8,417

To	*the city*	*the sea*	*the mountains*	*the countryside*	*Responses % of total*
From					
Austria	23.3	59.5	24.7	22.0	2.2
Belgium	30.1	58.1	28.9	23.3	2.7
Denmark	47.4	58.5	30.7	32.5	1.4
Finland	41.2	29.6	12.8	40.1	1.3
France	22.8	68.0	29.2	23.5	15.1
West Germany	*17.5*	*64.8*	*31.1*	*20.7*	*18.1*
East Germany	*20.7*	*52.1*	*38.1*	*23.2*	*4.3*
Germany (total)	18.2	62.0	32.7	21.2	
Greece	27.6	78.8	19.9	8.2	2.8
Ireland	41.4	58.4	13.6	28.8	1.1
Italy	18.6	67.2	20.3	6.3	15.9
Luxembourg	26.2	68.2	33.7	22.9	0.1
Netherlands	31.6	45.4	36.5	39.6	4.1
Portugal	22.2	60.5	11.9	23.0	2.6
Spain	32.5	59.9	20.8	15.9	10.6
Sweden	41.0	54.5	23.7	36.1	2.4
UK	24.9	67.0	16.4	36.1	15.0
Total	2,060	5,287	2,093	1,948	12,353
	16.7	42.8	16.9	15.8	100.0

Source: *The Europeans on Holidays 1997–1998*. Eurobarometer Survey 1997.

An analysis of the main criteria for choosing a holiday destination indicates more clearly the market share and the pull factors of urban tourism (Table 6.2). 'Places of historic and cultural interest, monuments and museums' were mentioned in 32 per cent of the responses in the Eurobarometer survey. When the factor 'Entertainment (cinema, discos, theme parks)', which can be considered as mainly urban attractions and accounted for another 18 per cent of the responses, is added, the pull factor of cities can be more clearly understood.

Due to a lack of statistical data, urban tourism flows have so far been largely underestimated. A recent large-scale survey (3,500 respondents) carried out in France concerning holiday patterns in urban tourism confirmed the growing importance of urban tourism for regional and local development and showed that urban tourism is less seasonal than other forms (INRETS 1996). Urban destinations have a particular appeal for younger people (under 35), of whom three quarters visit a city at least once a year. The percentage of the senior visitors (over 65) going to a city is about 40 per cent. The profile of the urban tourist, according to this French survey,

Table 6.2 Main criteria for holiday destination choice (relevant for urban tourism destinations) in per cent of the responses. N=8,417

	Places of historical and cultural interest Museums, monuments	*Entertainment cinema, disco, theme park etc.*
Austria	32.2	15.1
Belgium	28.7	16.3
Denmark	50.5	11.1
Finland	23.4	11.7
France	24.7	14.0
West Germany	*35.0*	*15.6*
East Germany	*38.3*	*18.7*
Germany (total)	35.7	16.2
Greece	27.6	40.2
Ireland	16.9	21.1
Italy	35.3	27.4
Luxembourg	22.4	20.8
Netherlands	39.1	13.6
Portugal	19.8	14.8
Spain	28.5	16.5
Sweden	40.2	18.7
UK	28.5	17.3
Total	2,662	1,525
	12.5	7.1

Source: *The Europeans on Holidays 1997–1998*. Eurobarometer Survey 1997.

indicates a clear predominance of young, higher-educated, bachelors and families of four. The weakest urban orientation in leisure patterns is found amongst the unemployed and retired population, older people, lower income groups, rural inhabitants and single persons.

A national survey (5,110 respondents) in the Netherlands (NIPO 1990) concluded that half of the respondents had visited a city during their holidays in the previous year. In the preceding three years there had been an average of 2.3 visits to a city in another country. This survey also identifies a high proportion of urban tourists in the age group between 18 and 34 and in the higher social classes.

Such surveys allow a better understanding of the attraction of cities for holiday makers. The distinction between cities of transit and destination cities is highly relevant in terms of tourism potential, as is length of stay. On average, urban visits are short, only a quarter of the visits last one week or more. There is little seasonal differentiation, the only factor of influence being the school holidays. The results of this type of survey underline the need for further empirical research, in particular into motivation patterns, in order to understand the actual pull factor of urban tourism destinations (Page 1995). Analysing the pattern of motives and activities is a most useful exercise and eventually an effective basis for urban tourism development plans.

Obviously there is a wide range of activities in which urban tourists take part. In the context of this chapter on heritage and cultural tourism, the question is to assess the relative importance of cultural activities, which include visits to heritage buildings and sites. A survey carried out in Europe by KPMG (1993) concerning images and pull factors of urban destinations concluded that the main pull factors are culture, shopping, sightseeing, events and sports. This is very much in line with the results of the French survey, in which it was stated that cultural activities account for 40 per cent of the activities of tourists in the city. In particular, first-time visitors focus on the cultural and historic aspects of the place, they go sightseeing and strongly follow the recommendations of guidebooks. The second group of activities is related to shopping (15 per cent of the activities) and attending events and entertainment facilities. The role of restaurants, pubs, attractions and night-life is most significant as a pull factor. Apparently there is also a small (5 per cent) but growing interest in industrial heritage and in social and industrial history (Prentice 1994).

A systematic comparison of the profile of urban tourists in different European countries is not yet possible; the surveys each have a different format, vary with regard to place and time dimensions and include different definitions of activities. This remains a serious handicap in the interpretation of international trends in urban tourism (Jansen-Verbeke 1996).

The Dutch survey (NIPO 1990), which dealt with visits to cities outside the home country, established the following ranking of motives for city visits:

1	Sightseeing	58%
2	Museums and other attractions	36%
3	Shopping	26%
4	Visit to friends and relatives	10%
5	Wining and dining	7%
6	Pub visit	5%
7	Transit	4%
8	Ambiance	3%

Urban tourism developers and marketers need this kind of information on motives and activities to target their plans for future development and marketing (Light and Prentice 1994).

A complementary approach consists of assessing the process of image building and the way this determines destination choice. In a comparative study of ten European urban destinations, KPMG (1993) came to the conclusion that the following factors determine the image of places:

1 Ambiance – Liveliness
2 Historic – Cultural heritage
3 Museums
4 Tourist attractions
5 Cultural activities
6 Price
7 Security
8 Restaurants
9 Public transport

From the results of the different surveys both the general and more specific pull factors for city trips can be identified (Figure 6.2). This overview holds strong arguments for focusing on cultural and heritage resources as principal components in the development of urban tourism, although specific information on the attraction scores of monuments, museums and other heritage features is not yet available. A second conclusion to be drawn from these surveys is that urban tourism cannot be regarded as an 'isolated attraction' of the city but is by definition strongly anchored in the urban morphology and the functional urban system (Murphy 1992).

THE SYMBIOSIS OF CULTURE, HERITAGE AND TOURISM

The current success in attracting tourists to urban destinations is strongly related to a 'cultural revival' in society and in tourism in particular. Cultural tourism is seen to be the fastest growing segment in the tourism market

General pull factors	Specific pull factors
1. Unique and interesting	Lots to see and to do An interesting place Unique experience
2. Cultural attractions and sightseeing	Well known landmarks Interesting architecture Noted for history Excellent museums and galleries Interesting local people Different culture and way of life Local customs and traditions
3. Entertainment	Exciting night life Exciting shopping Live music Theatre and arts Interesting festivals & events
4. Food and accommodation	Good hotels Sophisticated restaurants Typical cuisine

Figure 6.2 Pull factors for city trips.

(Richards 1996). A growing interest in cultural heritage, in historic places and cities offering a diverse cultural agenda, opens new perspectives for the urban economy (Van den Borg *et al.* 1996). This implies new challenges for the tourism industry, for the management of cultural facilities and for public agencies (Swarbrooke 1994). Urban authorities show a growing interest in cultural and heritage resources, with an emphasis on infrastructure which can eventually be developed as a motor for economic development.

The revalorization of European heritage is a new trend, strongly supported by EU action programmes (e.g. RAPHAEL). This includes an awareness programme concerning cultural values in Europe, the development of networks, the accessibility and reuse of heritage buildings, innovation and the professionalization of heritage management (EC, DG X 1997). The impact of these actions on urban tourism is considerable. The pilot project on 'Art Cities in Europe' established a network of forty-two cities in fifteen European countries, with the common objective of strengthening the symbiosis between cultural heritage and tourism by application of new information technology. The agenda of cultural events in European cities is also booming, which can be seen as another expression of the cultural revival and new market niches.

Synergy between culture and tourism

The mutual benefits for culture and tourism are driving both sectors towards defining common economic targets. The conservation of cultural resources and the process of transformation into tourism products can be a real incentive to the process of reviving cultural identity, on the community or regional level. In its turn this process creates a favourable incubation climate for the development of and investment in new tourism projects, which the tourism market needs in its current search for innovation and diversification. It is generally assumed that culture and tourism are interdependent (Ashworth 1993). Given the high costs involved in the conservation of cultural heritage and the exploitation of cultural facilities, tourism revenues are badly needed. The dynamism of culture, in its different forms and expressions, finds an incentive and in many cases genuine support in tourism. From the point of view of culture, tourism is also seen as a way of legitimizing political support, a social (and economic) justification, a means towards conservation and an incentive for innovation. In many countries tourism has become an important sponsor of cultural heritage for which there is no alternative. The tourism market needs cultural and heritage resources in order to develop new products. These products and facilities add value to the tourist experience, so the interests of both sectors, culture and tourism, are highly compatible.

According to many observers, this relationship of interdependency has even become inevitable. From the tourism point of view, culture, in its widest definition, is seen as a ubiquitous resource which can be developed into a tourist product (Ashworth 1995). To every city there is a history, every monument has its story – so the potential appears to be unlimited. Furthermore, the transformation process from cultural resources to tourist products is not necessarily a high-cost investment when compared to other tourism infrastructure projects. Probably one of the most relevant incentives to invest in cultural tourism products is the fact that there is no licence for cultural and heritage resources, often no ownership can be claimed; in many cases they are a public good, a common pool where every creative tourism developer can fish (Healy 1994). Many tourism entrepreneurs in a historic city benefit from the historic environment, without substantially paying for this added value to the tourist product.

Cultural and heritage resources in particular play a key role in the development of urban tourism. This strong belief in the potential of cultural tourism as a core business for urban tourism leads to a key question; is this view indeed supported by a structural change in the demand side? Academic researchers, tourism marketers and managers in the tourism business are challenged to understand better the characteristics of this 'cultural' market segment, the push factors on the demand side and the pull factors on the supply side.

Profile of the 'cultural tourist'

There still is no real consensus about the definition of cultural tourism and cultural tourists, and even less about the motives of heritage tourists. The fact that 'cultural tourism' accounts for increasing volumes of travel can be understood as a shift in the motivation pattern of travellers; understanding other cultures, gaining new perspectives in life, visiting cultural, historic heritage and archaeological treasures are now becoming important motives for travelling (Herbert 1995). In order to assess this trend either as a temporary and superficial shift in the market or possibly as a more fundamental sociocultural change we need to understand more about the hidden agenda of people coming to visit a city and pretending that 'sightseeing' is their prime motive (Jansen-Verbeke and Van Rekom 1996). Is there an indication of a genuine cultural interest or is the combination with shopping and other leisure activities actually a more realistic understanding of urban tourism (Jansen-Verbeke 1994)?

The key questions then are: who is this 'cultural tourist', what are the motives, the activities, the experiences and the characteristics of this market segment and, how important is this segment? It would be interesting to identify sharply this segment of the cultural tourist in the many tourist typologies which have been published so far. Obviously, market segments can be identified in many different ways, depending on the objectives of the study. For instance, the market segmentation and typology of European tourists, based on an international survey of life styles, hold many interesting clues (Mazanec and Zins 1994). To what extent are cultural motives and activities relevant factors in this market segmentation? Based on different typologies and survey studies, the market of cultural tourists can be divided into three distinct segments (Figure 6.3).

The culturally motivated tourists select a holiday destination as a function of their specific interests and the cultural facilities offered in a destination area. They are highly motivated to learn and to benefit from each opportunity. They will spend several days in a particular destination (city or

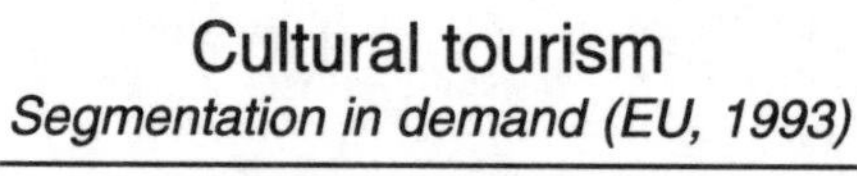

Figure 6.3 Cultural tourism.

region) and go around well prepared and with a professional tour guide. This 'ideal' cultural tourist only represents a small minority of the market, according to a European survey (Bywater 1993).

The culturally inspired tourists are attracted by special cultural themes, they visit the well-known places of culture, the major exhibitions and festivals, they travel around and collect experiences in many places, never stay long in one place. For this group of tourists, places like Venice, Athens, Canterbury Cathedral are on the agenda. According to several forecasts, a growing number of travellers will belong to this type of 'culture consumers' (Van der Borg *et al.* 1996).

For the culturally attracted tourists, while being on holiday in a seaside or mountain resort, an occasional visit to a city or historic site in the hinterland, to a museum, a church, a monument, is seen as a welcome diversion in the holiday programme. In fact they do not select their holiday destination on the basis of these facilities, but once there, they tend to appreciate greatly the opportunities available. Since the trend towards more active holidays is 'in', it is very likely that this pattern of cultural pastime will increase (Light and Prentice 1994).

Despite the common characteristics in motives and behaviour, there remains the question about the real motives of the individual. Motivation is the hidden agenda behind tourist decision processes and behaviour patterns.

Heritage resources and cultural tourism

Many historic cities in Europe have a rich heritage on which to build a tourism function (Van der Borg *et al.* 1996). This implies the need for a systematic evaluation of the way in which the cultural and heritage resources are anchored in the urban or regional tourist opportunity spectrum and of the trends and demands in the market of cultural and urban tourism.

In many urban tourism destinations there is at present a selection process involving heritage resources which can or could be developed into tourist products. In some respects, the process of developing tourist products from cultural and heritage resources can be compared with an industrial production process; there is a basic product, there is a transformation process and there is an end product which is on offer to the market. This is a simplified view of the tourism production process, since in fact the final tourist product is the 'experience by the tourist' in which many different actors and factors play a role (Urry 1990).

As Figure 6.4 shows, this production and selection process has at least two groups of stake holders. On the one hand there are the conservation agencies of cultural and heritage resources who hold their own values and select the artefacts and sites which, in their view, could benefit from tourism. On the other side there are the tourism marketers who evaluate the cultural and heritage resources from the point of view of tourism potential and marketing

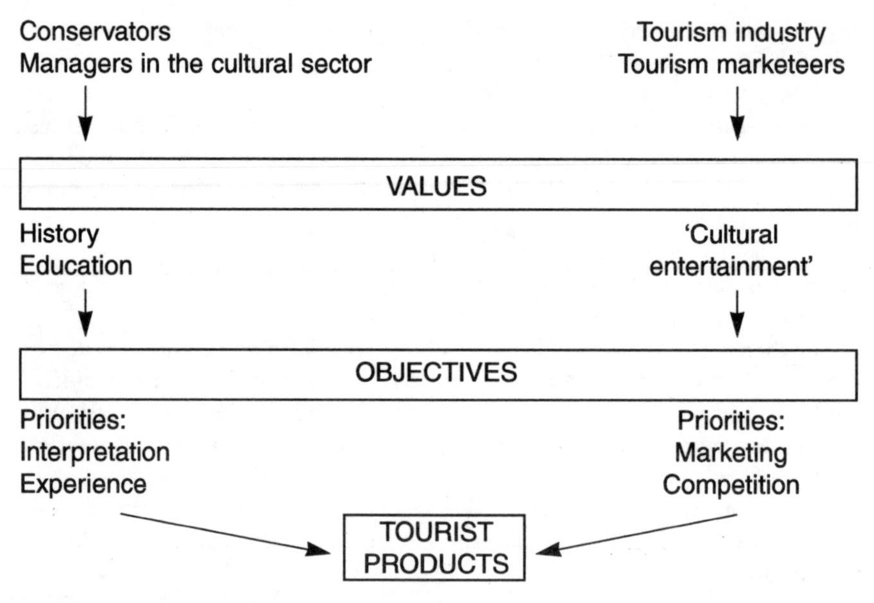

Figure 6.4 Selection of heritage resources.

investments. The selection process can lead to different and in some cases opposing views on the strategies of product development. In fact, there is a wide range of cultural and heritage resources which can be transformed into tourist products, which explains the importance of a selection procedure. In the process of developing and planning tourism products, the step of selecting the most interesting artefacts and sites and assigning a particular function and interpretation is crucial and often irreversible. Introducing heritage 'attractions' into the mental map of the tourist is indeed a first and decisive step in the process of 'touristification' of the historic city.

In every historic city entering the stage of tourism development, this discussion takes place, sometimes at the scale of urban districts, sometimes at the level of buildings and artefacts. As a rule there is a strong tendency to imitate successful examples taken from other cities, with a rather weak concern about the cultural identity of the place and a lack of understanding of what 'sense of place', the added value to the tourist experience, actually means. Arguments for a particular use are based on the apparent convergence in tourist preferences, expectations and behaviour, a 'standard image' of an attractive historic city strongly promoted by the media. This leads towards more imitation in the development of tourist products and in the way urban public space is being restructured for tourism purposes. Convergence on the demand side and standardization of products on the supply side fit well into the current views on and the practice of economies of scale. This also explains why the option of diversification and divergence in tourist product development, with an emphasis on uniqueness and cultural identity of resources, place and people, is regarded as the more 'risky' option.

However there are clear signs that gradually awareness grows regarding the competitive advantages of communicating cultural identity and the uniqueness of a place. In this search for the uniqueness of the place and for strategies to integrate heritage resources in the specific urban tourist opportunity spectrum, spatial analysis of the morphological and functional structure of historic cities opens most interesting perspectives.

SPATIAL ANALYSIS OF HERITAGE CLUSTERS

Introducing heritage buildings into the mental map of tourists and eventually into their activity space is not only a matter of place marketing. Prior to this step of promotion and communication, it is necessary to understand the urban tourist attraction system and the way heritage sites can become core elements of the urban tourist product. Several useful concepts about the urban tourist product, the tourist attraction system and the role of landmarks (MacCannell 1976; Leiper 1990; Pearce 1991) have been introduced earlier and are now being applied in field studies in Flemish cities in the research project, 'The role of cultural tourism in the revitalization of Flemish cities', being undertaken by the Institute for Social and Economic Geography of the Catholic University, Leuven. One of the objectives of the empirical study is to evaluate the capacity of cities to develop heritage resources into a core product of urban tourism. The tourism potential of historic cities is determined by a number of factors, of which obviously the presence of interesting heritage buildings is only one aspect.

Two themes are being explored: the first refers to the morphological characteristics of heritage buildings and their location pattern within the historic city. This 'form' aspect is seen to be of vital importance in structuring and developing attractive tourist zones and trails. The second concerns the integration of heritage buildings and complexes into the multifunctional structure of the city. The tourist attraction of heritage buildings strongly depends on their actual function and use and the way this is embedded into the urban tourist opportunity spectrum. The methodological approach and the empirical application of some experimental models will be briefly discussed, referring to the results of a pilot study in Leuven (Belgium) (Lievois 1998).

Morphological characteristics of heritage resources

The tourist attraction of a historic city or of specific urban districts depends, though not exclusively, on the characteristics of the built environment, on architecture, urban forms, artefacts and public spaces. A spatial concentration of heritage buildings and places in the historic city is assumed to have a higher tourist attraction than a range of single heritage buildings. This view is based on recent results of time–space analyses, which indicate the preferences of tourists for clustered opportunities (Dietvorst 1995).

An inventory of all 'classified' historic buildings, irrespective of their function or scale, forms the basis of the case studies. The lack of consensus on buildings with the status of 'heritage', or the criteria on which they have been selected for conservation, complicates this inventory. The criterion 'historic building' can be controversial, in particular when it concerns 'rebuilt' monuments and buildings in a traditional and vernacular architecture (Soane 1994).

For instance, the old market square of Leuven, in many ways appreciated as an interesting historic place in the city, is in fact completely rebuilt after being seriously damaged during the First World War. This raises the question about authenticity of the so-called heritage resources. Surely the actual age of the buildings is not really crucial to the appreciation of the average tourist (Urry 1990; Ashworth 1995).

The main purpose of the pilot study is a spatial analysis of the location pattern of heritage buildings and complexes. The analysis goes beyond the mapping of concentration areas by introducing the concept of clustering, based on walking distances between the different locations (measured along the street pattern). The model, comparable to a gravity model, includes a weighting of the buildings. The morphological positioning index takes into account the distance of buildings in relation to the landmarks. The method allows clusters or core areas to be identified: the higher the index the more central (distance wise) the building is situated in a cluster of heritage buildings. This method requires a selection of buildings or sites which can be considered as landmarks in the historic city which function as reference points in the mental map of urban tourists.

In the pilot study in Leuven, as nuclei for the clusters we have selected heritage buildings or sites, with a high marker value. In the external communication on tourist destinations reference is made to particular landmarks. The promotion material, targeted at the domestic market, was screened in order to define the hierarchy of landmarks for each city. The identification of the marker value of heritage buildings should also be deducted from the actual time–space behaviour pattern of tourists. In fact the validity of the selection was also checked from the tourist point of view (see pages 100–103).

The landmarks function as spatial reference points in the mental map of tourists and are therefore central to the analysis of the morphological positioning of heritage buildings and places of interest. The application of the network partitioning method in TRANSCAD makes it possible to define, along the street network, distance zones and clusters. The model is based on pedestrian visits to the historic city and marked different distance zones in the cluster partitioning. In the morphological positioning index several grades can be distinguished, depending on the number of markers, the marker value and the (walking) distances considered to be acceptable for tourists. This type of spatial analysis enables the positional strength of the different buildings in a cluster and hence the potential of the clusters within the historic city to be identified (Figure 6.5).

Clearly the tourist attraction of clusters increases with the degree of centrality, which in turn reflects the number, scale and marker value of heritage buildings in a particular area or street. This analytical model also allows a hierarchy amongst places of interest in the city to be established, based on the morphological characteristics and positioning. However, the positioning index is just one factor in the tourist attraction: the accessibility of heritage buildings for tourists, the tourist function and specific uses need to be evaluated as well.

Functional characteristics of heritage clusters

The tourist function of heritage buildings depends on the accessibility for tourists (e.g. public access and opening hours) and on the present use and

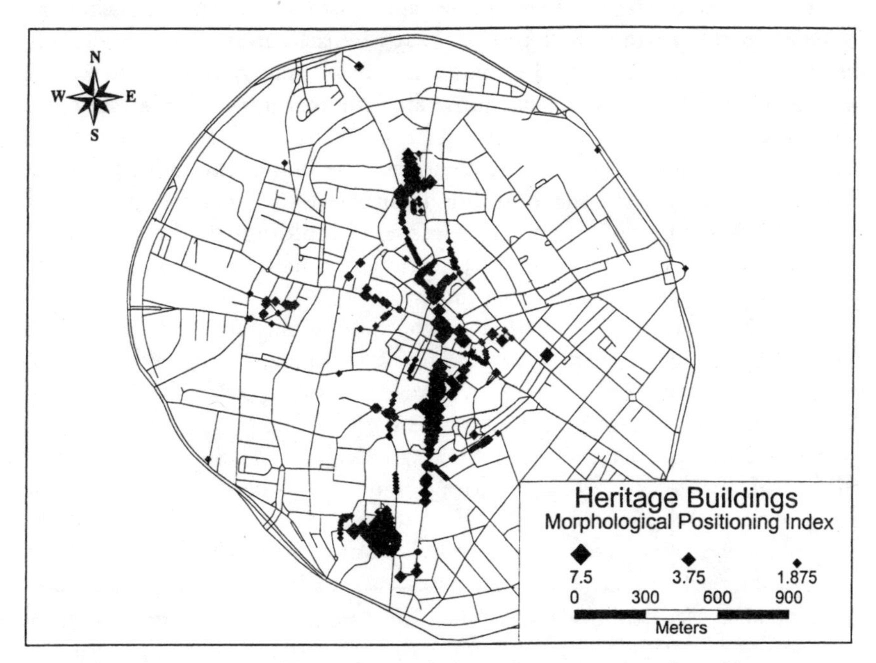

Figure 6.5 Historic city of Leuven: morphological positioning index of heritage buildings.

Source: Case study of Leuven, 1997, E. Lievois.

function of the building or complex of buildings. The latter can range from monofunctional (e.g. a visitor centre, a museum) to multifunctional (e.g. a combination of cultural activities, pubs, shopping). Heritage buildings can be open to the public and have an exclusive tourist function, or have a multifunctional use. In some cases they have no direct tourist use at all (e.g. residential buildings, university premises) but still add to the scenery and the historic setting of the city.

In the analysis of the tourist function of heritage clusters, the functional index refers to a combination of the accessibility for tourists and the range of uses/functions in the cluster. The highest functional score is given to buildings which have a core function in the tourist attraction, have a specific tourist function and are 'daily' accessible to tourists. This is based on an appreciation of the different tourist attractions in the city as core elements or secondary elements (Jansen-Verbeke 1994).

The conceptual model of tourist attraction applied in this analysis is a tool to identify heritage clusters and to rank these in terms of morphological positioning and of functionality (Figure 6.6). The results of this analysis in the historic city Leuven are illustrated by Figure 6.7.

The basic principle of this analytical model is a close interaction between form and function in the development of 'interesting' heritage clusters. There is a clear gradation between heritage clusters which function as core elements in the urban tourist product, both by their morphological characteristics and positioning and by their accessibility, and tourist function and heritage buildings which are not integrated, neither spatially nor function-

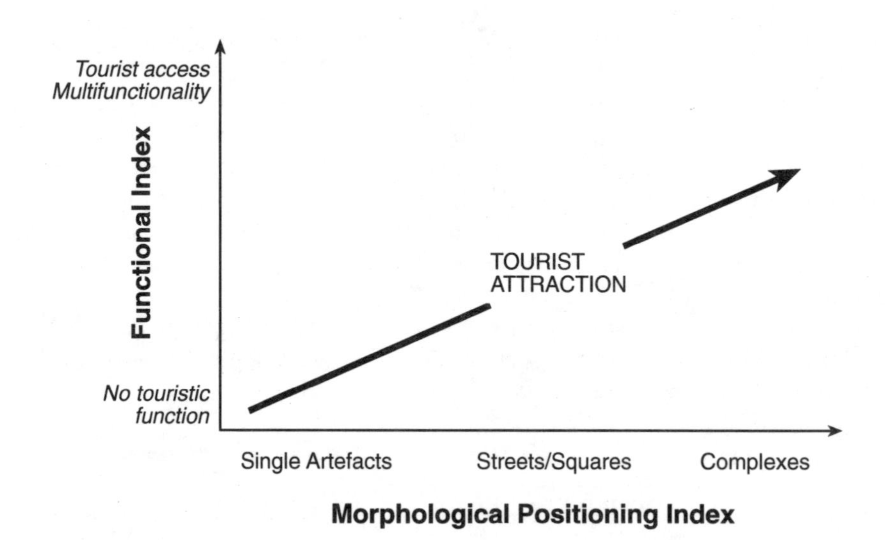

Figure 6.6 Tourist attraction of heritage clusters in a historic city.

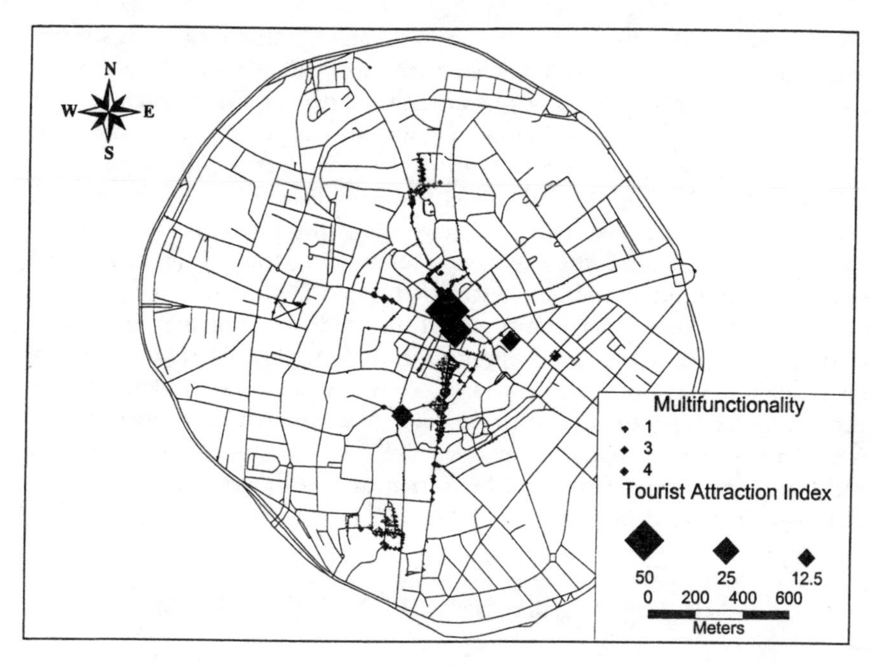

Figure 6.7 Historic city of Leuven: functional index and tourist attraction index of heritage clusters.

Source: Case study of Leuven, 1997, E. Lievois.

ally, in a cluster. The latter are less attractive for the average pedestrian tourist.

The first step in the analysis focused on the heritage clusters within the historic city without taking into account the overall functional structure and zoning in the inner city. The tourist function of a historic city cannot be seen as some implant in the urban fabric. On the contrary, it is part of the urban system in many ways (Ashworth and Tunbridge 1990). The growing importance of the tourist function gradually changes the functional mix in the inner city and the impact of tourism activities affects the environmental qualities and characteristics. The harmony and interaction between tourism and other economic activities, in terms of space use, strongly depend on the way in which the heritage clusters are physically and functionally integrated in the urban system.

Urban tourist opportunity spectrum

This view on spatial and functional integration of tourism into the urban system can be elaborated by referring to the tourist opportunity spectrum.

The concept is not new. Opportunity sets were introduced as analytical instruments for destination marketing, the Recreation Opportunity Spectrum (ROS) as a framework for planning, management and research (Clarke and Stankey 1979), and the Tourism Opportunity Spectrum as a context for understanding the interaction in tourism destination areas which meets the multiple needs of tourists, local residents, entrepreneurs and public agencies (Butler and Waldbrook 1991). These views on planning for tourism opportunities in natural resource based tourism development can be adapted to the context of urban tourism destinations.

The urban tourist opportunity spectrum needs to be defined in a more restricted way as the range of opportunities to which tourists have access, including a range of core elements and a diversity of secondary elements and supporting facilities which add to the value of the tourist experience (Figure 6.8). The concept of 'opportunity spectrum' from the tourist point of view and in an urban context needs to be further explored as a planning tool and as a marketing instrument.

URBAN TOURIST OPPORTUNITY SPECTRUM

=

ACCESSIBILITY TO A
RANGE OF URBAN FACILITIES and
ENVIRONMENTAL CHARACTERISTICS
of Potential Interest and Use for Tourists

Core elements
of the cultural tourism product

=

main pull factors for cultural tourists
high marker value

Heritage buildings and complexes ,
museums, monuments, events, theatres, etc.

+

Secondary elements
(added value to the tourist experience
and image)

Pubs, restaurants, street markets
and shops, etc.

Figure 6.8 Urban tourist opportunity spectrum: the concept.

The factors which play a role in the development of an urban tourist opportunity spectrum are:

- accessibility to and within the destination area;
- the possibility of choosing from a wide range of activities and meeting a diversity of preferences;
- the combination of activities within a specific time–space budget;
- the spatial arrangement of interesting places (networks, trails);
- the functional synergy between urban facilities;
- interaction between activities.

The results obtained from the spatial analysis of heritage clusters form an interesting framework for the study of the urban tourist opportunity spectrum. This implies a further analysis of the urban functions supporting tourism activities (e.g. pubs, restaurants, shops, events), the interaction and synergy between tourism and other urban activities and, not least, the activities and preferences of tourists in the city. A more detailed functional analysis of urban space use was part of the pilot study. A first interpretation of these results leads to defining three relevant dimensions in the development model of the urban tourist opportunity spectrum (Figure 6.9). The challenge now lies in a more in-depth analysis of each of these dimensions, in particular to assess the role of heritage clusters in the development of an attractive spectrum for different types of tourists.

The components of the tourist attraction index (combination of the morphological positioning index and the functionality index) offer a framework for the management of the urban tourist opportunity spectrum (Figure 6.10). In order to use this analytical approach effectively as a planning tool, much more information is needed on the components of the tourist

'Clustering'
Spatial concentration of complementary tourist products
Criterion: distances + images

'Synergy'
Multi functionality: Tourism + other urban functions
Criteria:
Compatibility of activities
Complementarity of functions

'Themed trials'
'Intervening opportunities'
Criterion: Seduction and surprises

Figure 6.9 Tourist opportunity spectrum development models.

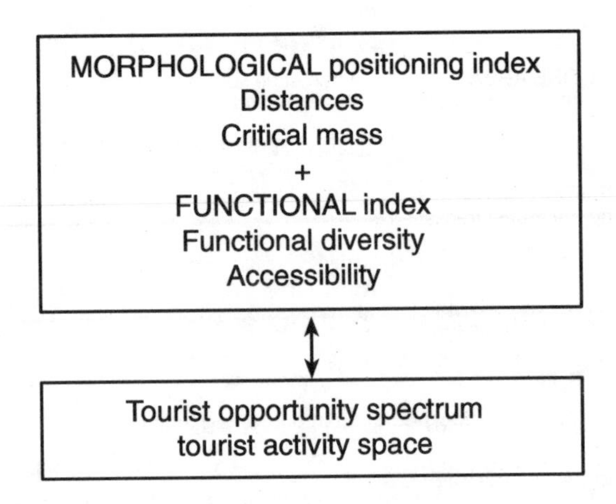

Figure 6.10 Components of the tourist attraction index.

attraction and on the activities and preferences of tourists. In particular the construction of themed trails and the role of intervening opportunities need further fieldwork.

Understanding the different components of the urban attraction for cultural tourists is not the final objective but a basis for the development of management strategies. These can be geared to increasing the tourism potential of the historic place, which is the case with newcomers in this market of cultural tourism destinations. The above concepts and models can also be implemented as planning tools in cities where tourism is reaching its capacity limits.

Assessing tourist attractions in an analytical and critical way opens new perspectives on sustainable development. Insight into the typology of heritage clusters, taking into account locational and functional characteristics, forms a solid basis for development options. Taking the example of Leuven, the options are for instance to strengthen the tourist function of heritage clusters which link the core areas, or to add tourist activities to the heritage sites which now function as sightseeing areas only (Figure 6.11).

TIME–SPACE USE OF URBAN TOURISTS

The spatial analysis of the historic city as discussed above takes into account the tourist point of view by referring to the marker value of heritage buildings or complexes as nuclei of the heritage clusters but it does not include as such an analysis of tourist action space (Pearce 1988, 1998). The concept of clustering is based on the assumption that pedestrian tourists discover the city within a limited time–space prism and have a mental map construct based on the presence of landmarks. The factors which determine

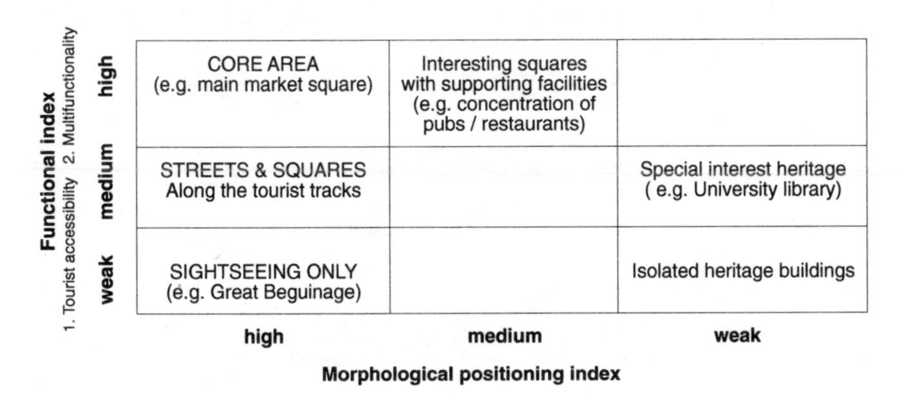

Figure 6.11 Typology of heritage clusters.

the urban tourist opportunity spectrum are equally based on the principle of combination possibilities.

Planning concepts such as the clustering of the urban tourist opportunity spectrum and networking (physical linkages between tourist clusters, functional and business alliances between suppliers and organizational partnerships between public and private partners) are now being developed and implemented in different places. This spatial approach to the process of 'touristification' needs to be completed and underpinned with the mapping of tourist action space (Figure 6.12). Research on time–space paths and budgets has a recently been given impetus through the different possibilities of applying GIS but work in this area remains rather experimental (Dietvorst 1995; Van der Knaap 1997).

An analysis of the time–space paths of tourists is therefore necessary, including an assessment of the appreciation scores (Figure 6.13). The pilot study in Leuven included an exploration of the motives, activities and appreciation of visitors to the historic city. The awareness of heritage clusters appeared to be a crucial factor in the way the city was discovered by tourists. This clearly points out the importance of promotion material and other communication media on which the expectations and images were built.

The role of the markers (which were used to construct the heritage clusters) proved to be highly significant for the routing (time and space use) of tourists. The survey carried out amongst visitors to Leuven was mainly intended as a methodological experiment on how to link data on time–space use with appreciation scores by applying network analysis (GIS/ TRANSCAD).

The tourist attraction of heritage buildings, as indicated by the above spatial analysis of clusters, was assessed from the tourist's point of view. This line of research is most promising in terms of marketing the best fit

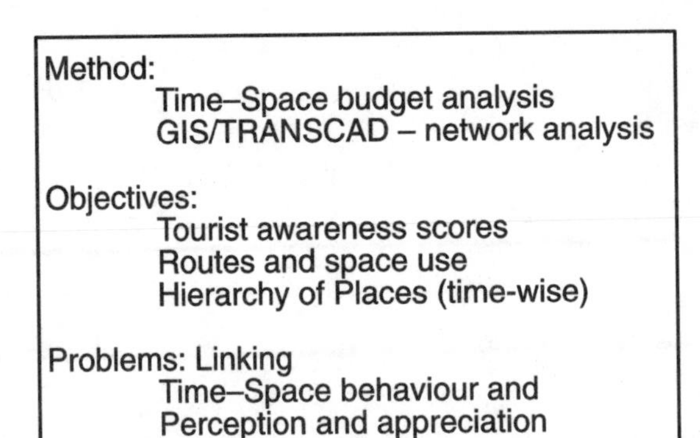

Figure 6.12 Tourist activity space.

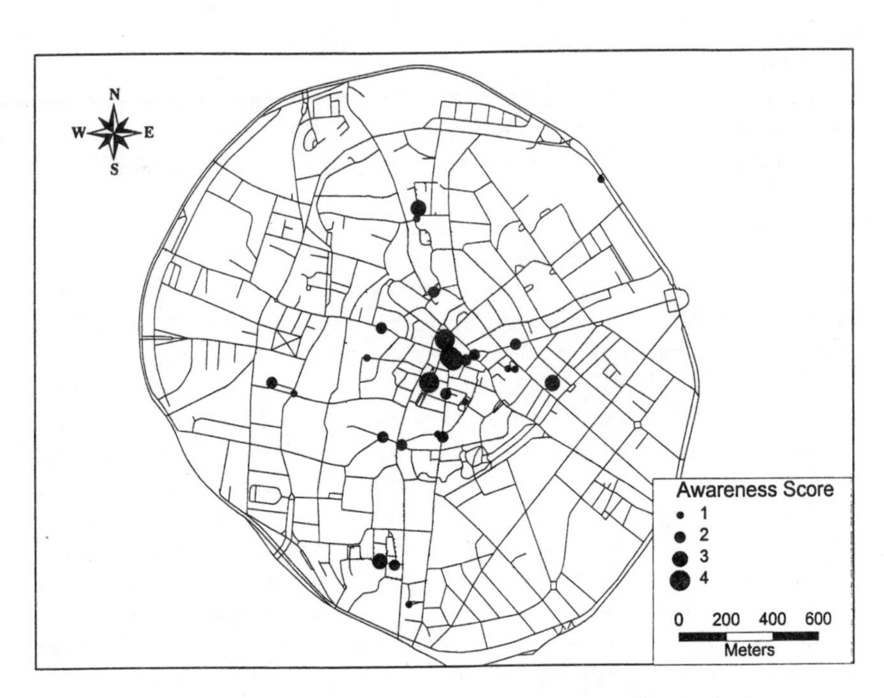

Figure 6.13 Historic city of Leuven: tourist awareness score of heritage clusters.

Source: Case sudy: Visitors in Leuven, 1998, Van Nunen.

place/product combinations and is also useful in indicating the weak spots in the functional pattern of tourist facilities. The validity of this approach to heritage clusters as a principal component in the urban tourism structure can be concluded from a comparison between the two means of analysis. A high congruence between the results concerning the tourist attraction index of heritage clusters was found (in 75 per cent of the cases), which is an incentive to proceed with this line of research. Some specific disparities are equally interesting; the appreciation of tourists concerning the heritage resources in a historic city tends to be based on areas (squares, streets) rather than on single buildings. Apparently, the mental map before the visit was strongly based on markers (single buildings), whereas the appreciation after the visit is mostly structured by clusters and a more general appreciation of the urban tourist opportunity spectrum.

RESEARCH AGENDA

Both lines of research need to be developed further, taking advantage of GIS applications in network analysis. In order to manage the urban tourist opportunity spectrum, a full understanding of all the factors is required. Insight into the tourist activity patterns (in time and space) is a first step in the process of developing strategies for visitor management and eventually too for the management of heritage resources. The urban tourist opportunity spectrum is not only an effective planning tool but also a framework in which the process of 'touristification' of historic cities can be understood and controlled. The field studies confirm the statement that the development of heritage tourism in historic cities initiates an irreversible process of 'touristification' in forms and in functions. This will indeed become a prime concern for urban managers and cultural conservators in coming years.

It cannot be denied that heritage tourism holds many opportunities for historic cities. However, strategic planning with a view to sustainability needs to be addressed at an early stage of the product development process (McNulty 1993). There are several threats involved in this transformation process from heritage resources to tourist product in an urban setting. Conflicts can be anticipated or even avoided by understanding the interaction patterns between tourist activities and the urban environment in which they are developed. It is inevitable that growing numbers of tourists and an increased spatial claim for tourism-related activities induces spatial transformation processes and pressure (Ashworth and Dietvorst 1995). Changes in the morphology of the place, in the physical structures, in the functional patterns, in the use of public space all contribute to this transformation process. In fact many aspects of daily life in the 'tourist city' are affected by the tourist activity. The impact of tourist activities is complex and one needs to understand the different ways in which these are changing the urban scene. New activities are induced by tourism development interfering with the existing system, thereby risking imbalance in the system (Jansen-Verbeke 1997). By analysing the different components of this urban dynamism, the unwanted effects of tourism can be anticipated and to some extent perhaps avoided.

Responsible planning is a key issue on the agenda of urban tourism planners, for which appropriate tools need to be further developed and tested in the field. Looking at tourism as an integrated function of the urban system may reduce the risks of imbalance between tourism, culture and other social functions (McNulty 1993). This implies an understanding of the complexity of the tourist system and of the changes in demand which now tend to multimotivation and as a consequence require a 'multifunctional' supply of facilities (Pearce 1991). In the context of a historic city, with a range of heritage clusters attractive to tourists, and by definition a multifunctional vocation, the management of particular resources can no longer be isolated from an integrated urban management policy. This 'new' management field is now in search of instruments and clear guidelines concerning regulations, implications, intended and unintended effects.

There is a growing consensus that tourism needs anticipatory policies and this reinforces the debate on responsibilities and tasks. However, the time horizon of the different agencies varies considerably; the tourism industry focuses on seasonal and yearly business reports, the local authorities have the election period as their time horizon and the conservationists tend to look backwards, sometimes forwards, without being much concerned (nor informed) about the dynamics of the tourism market. This difference in time horizon and also in values turns every tourism planning process into a series of compromises.

CONCLUSION

Introducing, in theory and in good practice, the principles of sustainability by emphasizing the irreversibility of the 'touristification' process in historic cities tends to be seen as a negative approach and yields insufficient response. The low response to idealistic options such as sustainable tourism can be understood, because tourism planners and academics lacked the expertise to translate the models and concepts into use and because examples of good practice are rare. Heritage tourism is now at an important crossroads: the mission to market nostalgia, authenticity, education and entertainment in such a way as to enrich the tourist experience and to safeguard the heritage resources for the future generations comes on top of the most important objective which is to use tourism as a stimulus for the urban economy and an added value to urban life.

References

Ashworth, G.J. (1993) 'Culture and tourism, conflict or symbiosis in Europe?', pp. 13–35 in W. Pompl and P. Lavery (eds) *Tourism in Europe*, London: Mansell.

—— (1995) 'Managing the cultural tourist', pp. 265–82 in G.J. Ashworth and A. Dietvorst (eds) *Tourism and Spatial Transformations: Implications for Policy and Planning*, Wallingford, UK: CAB International.

Ashworth G.J. and Dietvorst, A.G.J. (1995) *Tourism and Spatial Transformations: Implications for Policy and Planning*, Wallingford: CAB International.

Ashworth, G.J. and Tunbridge, J.E. (1990) *The Tourist-Historic City*, London: Belhaven.

Butler, R.W. and Waldbrook, L.A. (1991) 'A New Planning Tool: the Tourist Opportunity Spectrum', *Journal of Tourism Studies* 2 (1): 2–14.

Bywater, H. (1993) 'The market for cultural tourism in Europe', *EIU Travel & Tourism Analyst* 6: 30–46.

Cazes, G. and Potier, F. (1996) *Le Tourisme Urbain*, Paris: Presses Universitaires de France.

Clarke, R. and Stankey, G. (1979) *The Recreation Opportunity Spectrum: A Framework for Planning, Management and Research*, US: Department of Agriculture and Forest Service, Pacific Northwest Forest and Range Experiment Station, General Technical Report, PNW-98.

Cortie, C., Dekker, S., and Dignum, K. (1993) 'Prosperity and city types in the European Community', *Planologisch Nieuws* 13 (2): 117–26.

Dietvorst, A. (1995) 'Tourist behaviour and the importance of time–space analysis', pp. 163–82 in G.J. Ashworth and A. Dietvorst A. (eds) *Tourism and Spatial Transformations. Implications for Policy and Planning*, Wallingford, UK: CAB International.

European Commission, DG Regional Policy (1992) *Urbanisation and the functions of cities in the European Community*, Regional Development Studies, Luxembourg, 4.

—— DG Regional Policy and Cohesion (1997) *Community Involvement in Urban Regeneration: Added Value and Changing Values*, Regional Development Studies, Luxembourg, 27.

—— DG X (1997) *Information, Communication, Culture and Audiovisuals*, RAPHAEL Programme, 1997–2000.European Commission, DG XVI (1997) *Urban Pilot Projects Newsletter*, Winter 1997.

—— DG XXIII (1998) *Draft Summary Report on Tourism in Europe*, February 1998.

—— (1998) *The Europeans on Holidays 1997–98*, Eurobarometer Survey, Brussels.

Healy, R. (1994) 'The "common pool" problem in tourism landscapes', *Annals of Tourism Research* 21 (3): 596–611.

Herbert, D.T. (ed.) (1995) *Heritage, Tourism and Society*, London: Mansell.

Hinch, T.D. (1996) 'Urban Tourism: Perspectives on Sustainability', *The Journal of Sustainable Tourism*, 4 (2): 95–110.

INRETS (1996) *Le Tourisme Urbain – Les Pratiques des Français*, Rapport no. 208, Institut National de Recherche sur les Transports et leur Sécurité, Paris.

Jansen-Verbeke, M. (1994) 'Synergy between shopping and tourism; the Japanese experience', pp. 347–62 in W. Theobald (ed.) *Global Tourism; The Next Decade*, Butterworth–Heinemann.

—— (1996) 'Cultural tourism in the 21st century', *World Leisure and Recreation* 1: 6–11.

—— (1997) 'Urban Tourism: Managing Resources and Visitors', pp. 237–56 in S. Wahab and J. Pigram (eds), *Tourism, Development and Growth*, London and New York: Routledge.

—— (1998) 'Touristification of historic cities; a methodological exercise', *Annals of Tourism Research* 25 (2) (forthcoming).

Jansen-Verbeke, M. and Van de Wiel, E. (1995) 'Tourism planning in urban revitalisation projects', pp. 129–45 in G.J. Ashworth and A. Dietvorst (eds) *Tourism and Spatial Transformations*, A. Wallingford UK: CAB International.

Jansen-Verbeke, M. and Van Rekom, J. (1996) 'Scanning museum visitors. Urban tourism marketing', *Annals of Tourism Research* 23 (2): 364–75.

KPMG (1993) *Toeristische Concurrentiepositie van Amsterdam ten opzichte van tien andere Europese steden,* Amsterdam: Klynveld Management Consultants.

Law, C. (1992) 'Urban tourism and its contribution to economic regeneration', *Urban Studies* 29 (3/4): 599–618.

—— (1993) *Urban Tourism: Attracting Visitors to Large Cities*, Tourism, Leisure and Recreation Series, London: Mansell.

Leiper, N. (1990) 'Tourist attraction systems', *Annals of Tourism Research* 17(3): 367–84.

Lievois, E. (1998) 'De toeristische functie van historische gebouwen in de stad Leuven: een ruimtelijke analyse', in *De Aardrijkskunde* (forthcoming).

Light, D. and Prentice, R. (1994) 'Market based product development in heritage tourism', *Tourism Management* 15 (1): 27–36.

MacCannell, D. (1976) *The Tourist: A New Theory of the Leisure Class*, New York: The Macmillan Press Ltd.

McNulty, R. (1993) 'Cultural tourism and sustainable development', *World Travel and Tourism Review* 156–62.

Mazanec, J. and Zins, A.H. (1994) 'Tourist behaviour and the new European life style typology', pp. 199–216 in W. Theobald (ed.) *Global Tourism: The Next Decade*, Oxford: Butterworth–Heinemann.

Murphy, P. (1992), 'Urban tourism and visitor behaviour', *American Behavioural Scientist* 36 (2): 200–11.

NIPO (1990) *Survey of City Trips of the Dutch Population*, Amsterdam: Nederlands Instituut Publieksonderzoek.

Page, S. (1995) *Urban Tourism*, London: Routledge.

Pearce, D.G. (1988) 'Tourist time-budgets', *Annals of Tourism Research* 15 (1): 106–21.

Pearce, D.G. (1998) 'Tourist districts in Paris: structure and functions', *Tourism Management* 19 (1): 49–65.

Pearce, P.L. (1991) 'Analysing tourist attractions', *The Journal of Tourism Studies* 2: 46–55.

Prentice, R. (1994) *Tourism and Heritage Places*, London: Routledge.

Richards, G. (ed.) (1996) *Cultural Tourism in Europe*, Wallingford, UK: CAB International.

Soane, J. (1994) 'The renaissance of cultural vernaculism in Germany', pp. 159–77 in G.J. Ashworth and P.J. Larkham (eds) *Building a New Heritage*, London: Routledge.

Swarbrooke, J. (1994) 'The future of the past: heritage tourism into the 21st century', pp. 222–9 in A.V. Seaton *et al.* (eds) *Tourism: the State of the Art*, Chichester: John Wiley & Sons.

Urry, J. (1990) *The Tourist Gaze, Leisure and Travel in Contemporary Societies*, London: Sage Publications.

Van den Borg, J., Costa, P., and Gotti, G. (1996) 'Tourism in European heritage cities', *Annals of Tourism Research* 23 (2): 306–21.

Van der Knaap, W. (1997) 'The tourist's drives', PhD thesis, University of Wageningen, Wageningen.

7

THE POLITICS OF HERITAGE TOURISM DEVELOPMENT

Emerging issues for the new millennium

Linda K. Richter

'How much remembering is the right amount?'
(Horn 1997: 60)

INTRODUCTION

Heritage tourism has become a rather elastic term applied by some to almost anything about the past that can be visited. Such tourism may involve museums, historic districts, re-enactments of historical events, statues, monuments and shrines.

Perhaps because the word 'heritage' sounds lofty and important, there may be a general assumption that such tourism is by definition good and its development uncontroversial. That would be wrong. All the motivations, expectations, problems and negotiations that surround any form of tourism development must be factored into the development of heritage sites. The desire for economic growth, jobs, tax revenues, civic pride and private benefits are all a part of the struggle for heritage development. As such heritage tourism is in no way immune from the battle over power and resources. Establishing a heritage district in some cities in the world is scarcely easier than creating a red-light district!

However, there are political issues and challenges that are more central to heritage tourism development than other forms of tourism. Heritage tourism would appear to be a potentially important form of development, for example, as a source of national identity, political communication and socialization; but there has been little empirical research done on this topic. This chapter suggests where such research is needed. It also illustrates how this expanding form of tourism has become a growing arena for political conflict.

The issues of heritage development are central to what Harold Lasswell called politics: 'Who Gets What, When and How?' (Lasswell 1936). Heritage tourism also goes to the core of what constitutes our collective political memory, our national identity. As Garry Wills observes:

> If any culture is to be understood, historians must read it by judging what it chose to honour. The ancient Greeks did not praise warriors alone. They honoured philosophers. Medieval statues were of saints. The Renaissance brought classical gods back down to earth. Though we Americans do not think of ourselves as militaristic, the outdoor statues in Washington are overwhelmingly of generals. Is that an unwitting self-revelation, a kind of 'Freudian slip' in stone?
>
> (Wills 1997: 21)

Benign, militaristic, jingoistic, or 'Disneyfied', the heritage tourist experience may also help form attitudes and values that shape individual orientations toward groups, institutions and issues. Even the very substance of a heritage is a political construction of what is remembered – different for many groups in a society.

Ordinarily social scientists do not think of tourism as political socialization and communication because it takes place largely outside the familiar institutions of politics: the home, church, school, media and government. Heritage destinations may convey particular political messages – intended and unintended – that have been only rarely studied.

Because groups and communities are increasingly aware of the symbolic importance of being represented in heritage sites, interest group activity has concentrated on issues of what gets saved, destroyed, interpreted. Public policy has also revolved around decisions to commemorate political and historical legacies. In this chapter four emerging and interrelated political issues associated with heritage tourism will be examined. The first section looks at the changing balance of power in the preservation, development and presentation of heritage tourism. The second section details the growing number of claimants to political representation at heritage sites, while the third discusses the increased competition over issues of authenticity and interpretation. The fourth section explores the emerging popularity of commemorating tragic political events. At each juncture research gaps will be noted.

THE CHANGING BALANCE OF POWER

The struggle over heritage tourism reflects its growth and success. Everyone wants a piece of the action. To understand why the struggle has intensified one needs only to look at the magnitude of tourism involved.

Tourism is one of the world's largest industries. Solely in terms of international tourism, there were 613 million tourist arrivals in 1997 and international tourism receipts amounted to US$448 billion (WTO 1998). Heritage tourism's growth, while trickier to measure, is also substantial. For example the number of registered historical sites in the United States went from 1,200 in 1968 to 37,000 by 1987. A new museum is opening every week. Currently the United States has over 8,000 museums with over one billion visitors a year (Horn 1997: 54). In the United Kingdom the number of heritage sites has grown to 500,000 historic buildings, 5,500 conservation areas and 12,000 museum (Dann 1994: 55).

Why the growth? Several explanations have been offered. Dean MacCannell attributes it to the alienation many feel from their stressed and busy lives (Graburn 1995: 167). John Urry adds it is 'clearly part of a process by which the past has become much more highly valued in comparison with the present and the future' (Dann 1994: 60–1). Maybe, but does that explain holocaust, internment and slavery museums? More likely from this writer's perspective, the museums offer a multifaceted look at a past neither known nor understood. The humanities are increasingly neglected in our schools and colleges at the very time tourists are seeking out heritage sites.

A fourth factor that may be involved is community pressure to expand the number of attractions and destinations to accommodate the huge numbers of tourists. Many sites are at capacity most of the year. Controlling growth in one place while promoting alternative cultural sites is becoming crucial for many governments. Like the most popular national parks, the premier cultural sites are being 'loved to death'.

Fifth, demographics also favour the growth of heritage tourism. The 'greying' of the tourist trade augers well for heritage sites. There is the expectation that older tourists may seek out such sites in disproportionate numbers (Dann 1994: 60).

Given the reasons advanced for the continued increases in heritage tourism, how has the balance of power shifted during this process? To answer that it is important to look at the overall context in which heritage tourism exists. Tourism has long been noted for its asymmetries in impact. Some of these disparities have been discussed in centre–periphery terms (Hoivik and Heiberg 1980) where some nations are tourist-generating nations and others primarily destinations. Tourism has also reflected imbalances in racial and gender hiring practices with minorities and women comprising a disproportionate share of the tourism jobs, but largely those at the base of the career ladder (Fredericks 1992: 12; Edgell 1993: 18; Richter 1994).

Heritage tourism is no different. Historically these sites have reflected dominant establishment views of what is important to remember and commemorate. That power is to be expected in the process of what gets saved has been ably illustrated by Donald Horne. In his splendid study, *The Great Museum*, he examines how certain figures often of equal importance in the history of several countries may be revered in one nation with many statues and museums and ignored in another (Horne 1984). What gets destroyed is often a deliberate political act as well. Often one of the first acts of revolution is destruction of the artefacts of the past (Ashworth and Tunbridge 1990: 255).

Conservative critics of the proliferation of new museums and/or the more controversial interpretations attached to museum narratives may cry 'political correctness', (Leo 1994: 21) but *revisionist* history is a time-honoured tradition as power shifts. Heritage tourism is not a stranger to class struggle! The current period following the Cold War has launched an unusual amount of revisionist history, joined as it is to democratization and anti-socialist processes world-wide. What conservatives overlook is that it is the political left, Lenin and Marx, that are being 'downsized' or forgotten and new figures are emerging centre stage even in the museums and monuments. It was not very long ago, however, that museums were repositories for the glory and loot of the empire. In fact the creation of some British museums was explicitly designed to instill working class awe of 'their betters'.

Friedrich Nietzsche would probably have approved of such didactic heritage tourism: 'How could history serve life better than by tying even less favoured generations and populations to their homeland and its customs.' On the other hand, he would have sympathized with policymakers on tight budgets faced with demands for funds to preserve something dear to the heart of their constituency. 'The fact that something has become old now gives rise to the demand that it be immortal' (Nietzsche 1988: 20–1).

If the process of revisionist heritage tourism is not new, what *is* new is first the number of attractions designed to remember the *marginalized groups* in society. Today many of the new monuments and museums reflect groups that even today get only passing attention in mainstream history books.

Secondly, decisions may increasingly be divorced from the individuals who will be affected by the site. As heritage tourism is taken more seriously the number of political actors who function as veto groups or facilitators of projects has proliferated. Political leadership and indeed national and international organizations as well as foundations and associations are becoming seriously involved in the issues of funding, design and interpretation of historic sites. UNESCO's involvement in the restoration of Borabadur Temple in Indonesia and the historic city of Dubrovnik in Croatia, the Pacific Area Travel Association's heritage awards, and the US Congressional concerns over the Vietnam Wall, the Enola Gay, and the Franklin Roosevelt memorial suggest the variety of issues involved. Specific decision-making roles attached to this commitment by outside groups, and their consequences for those who are affected by their involvement have not been studied.

A third factor in the changing balance of power is the global fascination with downsizing the public sphere while increasing privatization of goods and services (Richter and Richter 1995). Heritage tourism has also reflected this trend, becoming more entrepreneurial and entertainment-oriented in an effort to compete with theme parks even as the latter try to look more educational (Graburn 1995: 167). Research needs to explore how the change in funding is impacting what gets remembered and how. Will we someday see the Coca Cola wing of the British Museum or the Toyota Deer Park in Nara?

Fourth, along with the broadening of museum or heritage tourism subjects and changes in funding, there has also been a democratization of the process by which tourists absorb the experience. 'Museums . . . are now no longer viewed simply as showcases containing trophies of a bygone era. They are to be entered and possessed' (Dann 1994: 66). Thus, when one tours the Holocaust Museum in Washington DC one has the option of taking a card with the name and description of a holocaust victim. More information is added at intervals in the tour until the details of the person's death are noted. The intent is obviously to build empathy and infuse reality into the museum's horrific story.

Moreover, the museums and sites are themselves being transformed by the tourist. Beyond the comments section of the guest books, a new expectation of participation is being forged between the tourist and the heritage site, with either able to initiate the communication.

How did this happen? No one has a definitive answer, but some speculate that it was influenced in the United States by the powerful emotions evoked by the AIDS quilt and the Vietnam Memorial:

> The most dramatic transformations in museums is their inclusion of voices once disqualified as too lowly, unscientific, and naive. The AIDS quilt like the teddy bears and poems left by visitors at the Vietnam Veterans Memorial . . . demonstrated how ordinary people could help shape the places that preserve the nation's memories.
>
> (Horn 1997: 60)

> . . . these new museums have remade themselves into true public forums, places for communal experience and exchange, for sorting out group and national identity. The histories they tell are no longer authored by expert curators alone. They incorporate testimony from ordinary people, witnesses, participants even visitors to the museum.
>
> (*ibid*: 54)

In some museums tourists are invited to add information they may have about individuals featured in the museum. Not all information and advice is enthusiastically received. It was public pressure in April 1997 by disabled people, for example, that led to the FDR memorial being altered to reflect the former American president's disability.

Most museums are increasingly interactive in format. At the Museum of Tolerance in Los Angeles, it is possible at several points to determine how much or how little explanation is desired. Most unnerving to this writer was one computer terminal that allowed one to punch in one's state and city. Flashed onto the screen are the names and numbers of hate groups in that area!

To generations accustomed to television and the computer, the new museums offer an accessibility not based exclusively on lecture and printed captions. Moreover, with the Internet, one can redefine what it means to be a tourist, visiting and interacting with sites transported by modem to another era and another world.

Finally, there may be a changing balance of power that has not been acknowledged. Is the modern museum a Big Brother where an Orwellian level of control is created? Light, sound, music, even smells and temperature can be manipulated to affect the emotional intensity of the experience. People are often moved through the museum at a controlled pace, sitting, standing and listening according to a carefully choreographed scripting of the visit.

Would it not it be an ironic twist of fate if the techniques that allowed tourists to maximize their appreciation of these heritage sites were the same techniques that controlled and manipulated the visitor as never before?

NEW CLAIMANTS TO POLITICAL REPRESENTATION

'Of course, who gets to tell the story is the battle of the day.'
(Horn 1997: 60)

The new claimants to political representation in our memorialized past have at least two objectives: (1) to be in the story and (2) to have a say in how the larger story is told. Ten years ago, it would have been relatively easy to identify all the sites remembering minorities' and women's contributions to US history. That is no longer true. The civil rights movements of the 1960s and 1970s may have stalled, but their effect has been to recover much of the history neglected by the mainstream historians. Now that history is being put on display in scores of museums and statues, exhibits, and special events.

One of the most impressive examples is not strictly a political heritage museum, but an art museum exclusively featuring the work of women. In the process of acquainting visitors with the unfamiliar names and pieces, like puzzle pieces the captions reveal the neglect of women artists. 'The history of all times, and of today especially teaches that . . . women will be forgotten if they forget to think about themselves' (Louise Otto-Peters 1849). This quotation hangs as a cautionary tale in the National Museum for Women in the Arts in Washington DC. Her advice was taken when in 1987 this museum became the first in the world dedicated to the work of women artists. Many of those featured are acknowledged by their peers to be superior to the male artists – often brothers, sons and husbands – now famous in major world galleries. Some, like the women of the cloister who worked on the Bayeaux Tapestry in the eleventh century, remain unknown.

This museum is an example of how private resources were needed to address the fact that public galleries routinely overlooked women artists. In fact, some of the individuals who were renowned in their own time had been lost to the modern era until this museum opened. As the court painter to Marie Antoinette wrote in 1835 about the women painters and sculptors: 'It is difficult to convey an idea of the urbanity, the graceful ease, in a word the affability of manners which made their charm of Parisian society 40 years ago. The women reigned then: the Revolution dethroned them' (Personal tour 1993). In 1994 the museum featured an exhibit of art by Arab women which broke through the stereotypes of what these women are like and their artistic range ('Lifting the Veil' 1994). How will such art affect attitudes toward women and the Arab world? This is a question worth investigating.

The Buffalo Soldiers monument, dedicated in 1992, is an example of a recent effort to commemorate black soldiers who have fought in every American war but have seldom been recognized. The buffalo soldiers were men drawn into cavalry regiments and infantry units in 1866 from among the ranks of the US Coloured Troops. These men had served or were serving as volunteers after the Civil War

Other commemorative places are being developed that will tell the often unsavoury history of race relations in the United States. In Savannah, Georgia, for example, it is now possible to take tours centred around important sites in black history. Museums and markers detail the life of black abolitionist Harriet Tubman and the Underground Railroad by which she smuggled slaves to the North.

That this can happen in the American South reflects real changes in black power even as some state capitols still insist on flying the flag of the Confederacy! How secure are these new attractions? Who visits these sites? Do the races differ in what they take from these landmarks? Are the major tours or school field trips visiting these sites? We should find out.

THE INCREASING STRUGGLE OVER AUTHENTICITY AND INTERPRETATION

The politics of heritage tourism development seems fairly straightforward when planners seek to restore a historical site, build a monument, or combine entertainment with a celebration of cultural richness. The process is seldom so uncomplicated. The political climate has expanded the number of groups and interests that want to determine how that story is told. That poses problems because revisionist history may or may not be more accurate (Bruner 1993). At the very least it adds new intellectual uncertainties to what were once portrayed as unquestioned truths.

The Arizona Memorial in Pearl Harbour is a case in point. It has been interesting to see the transformation of this museum over thirty years, in part, because of the changing political and economic realities of Hawaiian tourism. The facts on which the Memorial is based are not in dispute, but the interpretation has become more delicately nuanced. The 'sneak Jap attack' of 7 December 1941, which President Franklin Roosevelt told the nation was 'a day that would live in infamy' is now recalled at Pearl Harbour as the tragic site in which the national interests of two great world powers collided. The films that describe that fateful day have gradually become less indignant and hostile as Japanese tourists to Hawaii have increased. It would be easy to claim there has been a whitewash. But it may be, on balance, a more accurate telling of events than the earlier accounts. It is important to consider how the racial and gendered organization of our memories affects both tourism and our political impressions of what we see.

Has there been a similar change in the Hiroshima Museum in Japan? Far fewer American tourists visit Hiroshima than Japanese visit Hawaii.

> That may change, however, if city planners get their way. There is an effort to redesign the city space, not simply in order to erase its dark brooding memory in the interests of prosperity but to enclose or isolate it. The effect, however, is to 'celebrate peace in its weightlessness' without the anger and pain and inevitably to transform Hiroshima's historical meaning.
>
> (Duara 1997: 142)

Meanwhile, the American Congress recently showed a decided unwillingness to reconsider the decision to drop the atom bomb on Hiroshima. The planned Smithsonian exhibit of Enola Gay, the airplane that dropped the bomb, occasioned harsh attacks because of the exhibit's interpretation of that event. Eventually, the issue became part of a larger battle over funding for the Smithsonian leading to the resignation of one curator and the Enola Gay being shown without any explanation (Holloway 1995: 32–3; Washburn 1995: 40–9). Once more there is irony in the fact that conservative politicians were more than ready to criticize 'socialist realism' in its control of the arts in the former Soviet Union, but apparently saw no parallel in their own efforts to censor and control.

The Australian War Museum also furnishes an example of the political subjectivity of exhibits. No curator by 1987 had been able to decide on a representation of Australia's involvement in Vietnam that could garner public support. Was it a noble effort to defeat communism and assist an ally, or a bloody blunder? (Personal interview March 1987).

Perhaps it is as well the United States made no such effort, but simply acknowledged its terrible cost by building the Vietnam Wall. Given the immense popularity of the wall today, it is easy to forget that this memorial was the subject of harsh attack from those who wanted a more celebratory memorial (Wagner-Pacifici and Schwartz 1991: 376–420). The fact that the design was done by a young Asian female only compounded the fear that it was a subversive propaganda statement ('Maya Lin: A strong, clear vision' PBS 1996). What actually happened was that individuals focused on the people who sacrificed instead of the disputed issues of what was at stake.

Popular as the wall is, additional sculptures have been added as groups fought for recognition of their role in this divisive struggle. The Vietnam Women's Memorial was added on Memorial Day in 1993. But before that, it went on a travelling exhibit. One would not think a bronze memorial on a flatbed truck in a mall parking lot in Wichita, Kansas, could occasion strong emotions. However, having seen it this writer can testify to the depth of emotions that ran through the crowd of several hundred. Vietnam vets kissed and hugged some two dozen Vietnam nurses, gave them flowers, reminisced about the women who had nursed them through long nights. Everyone there appeared to realize that whatever side they had been on during the war, it was a beautiful and overdue recognition to see that the nurses – eleven of whom had died serving – at long last were remembered. How long did those intense feelings last? What did it mean to male and female veterans? What was learned? What was remembered?

Indeed, the whole process by which heritage tourism has begun to include more and more groups is barely examined. For example, many mainstream museums are getting more attentive to the inclusion of women and minorities in their exhibits. Why is this so? Several explanations are possible. First, the increased clout of blacks, women, even gays has helped these groups to negotiate their convention and tourism decisions with increased attention from convention planners and tourism officials. For instance, the African-American convention business is growing rapidly. Increasingly, black convention planners have told cities their destination decisions will be based in part on the way black heritage is presented, the degree of black political representation in the community, and the political reception their members receive. One third of all corporate convention planners, in general, would boycott certain cities because of social issues (Eisenstodt 1995: 15).

Secondly, there appears to be more acceptance of the notion that the mainstream story has been too narrowly focused. A third possible explanation is related to the first, to broaden the clientele for the attraction. As minorities become more powerful and affluent, they become not only subjects but consumers of heritage tourism.

The answer may differ from setting to setting, but it is clear that the formula of merely adding or stirring in new groups in the mainstream story will not suffice if the group has much clout. For example, Colonial Williamsburg, faulted for 'sanitizing history' by glossing over slaveholding, fared no better when it added a mock slave auction. The black civil rights group, the NAACP, was furious, charging the auction trivialized the real suffering slaves had known. Nor were black employees of the Library of Congress impressed with the exhibit on slavery the Library had developed (Horn 1997: 56). In some cases it may be more infuriating to be included if one finds the interpretation unacceptable.

Still, not everyone is enamoured with the interpretations minorities have placed on heritage sites within their control. Complaints surfaced that the National Museum of the American Indian was giving too much attention to settlers' brutality and far too little to Indian violence. The Museum of African American History in Detroit has been criticized for referring to the 'rebellions of 1968' which to the mainstream media had always been known as 'race riots' (Horn 1997: 56–60). A 'rebellion' implies considerably more political legitimacy than a 'riot', but few who fought to restore order were likely to see the events in the same category as the Boston Tea Party!

The US National Park Service has tried to balance the interpretation at its historical sites. That is often difficult to do. Complex, ambiguous stories are less easy to tell or to remember. In some cases there is even rivalry within a marginalized group for recognition of their history. The National Park Service, for example, was encouraged to balance its explanation of the Crow victory at Little Big Horn, known as Custer's Last Stand, with attention to Sioux and Cheyenne involvement (Horn 1997: 56).

Another controversy that illustrates that no good deed goes unpunished was the belated Congressional decision in 1997 to bring a large suffragette sculpture out of the basement of the Capitol where it had languished seventy-five years. Supporters in Congress who had fought to have the sculpture moved to the Rotunda were stunned when black civil rights groups found the sculpture unacceptable because it featured three white women – Susan B. Anthony, Elizabeth Cady Stanton, and Lucretia Mott – but had overlooked the eloquent black women's rights crusader of the same period, Sojourner Truth. There were even suggestions that her likeness be chiseled on the sculpture. In this instance the sculpture was moved unaltered into the Rotunda, but in the future there will be greater Congressional caution, if not sensitivity, to the politics of heritage. Given the possibilities for conflict, some museums and national parks opt to present exhibits with several interpretations attached.

Most nations, however, are quick to commemorate the legacy of those in power. Since 1989 in Eastern Europe and the former Soviet Union, heritage sites are in fundamental transition as the legacy of Marxism is erased. Busts of Lenin and even the embalmed leader himself appear to be on the way out. The Museum of the Revolution in Cuba still exists (Seaton 1996: 242).

Probably the museums most susceptible to politicization are the Presidential Libraries and Museums. First, the person for whom the museum is named is often still alive when it is built. Secondly, those giving to the museum are generally those who are enthusiastic about the president being immortalized. One only has to tour the Nixon Library to get the impression that most of Nixon's two terms were devoted to foreign policy before he was forced to resign because a third-rate burglary sabotaged his presidency. More telling details of the issues leading to possible impeachment are downplayed.

Yet, it is the presidential libraries that often provide for a more accurate assessment of a presidential administration than that available from news accounts of the period. As the repository of classified information, they provide additional details which, once declassified, have altered the historical standing of such presidents as Truman and Eisenhower. Still, presidential museums and libraries perhaps say more about Americans than they do about the presidents. For all their ridicule of 'cults of personality', Americans may be unique in so honouring modern chief executives with their very own tourist attractions!

Authenticity issues

Authenticity is also an issue in the politics of heritage tourism. Is authenticity essential? And if so, in what form, from what period and by whose standards? Some tourist sites are more interesting as ruins than they would be if everything were rebuilt. Consider Pompeii, Byblos, Mohenjodaro or even some contemporary abandoned towns. Local residents may not like having their community called a 'ghost town', but it may be a more successful 'ghost town' than a nondescript unincorporated cluster of renovated houses.

> Restoration also raises issues of authenticity. Tastes change. Would we admire the Parthenon if it still had a roof, and no longer appealed to the modern stereotype taste . . . If we repainted it in its original red, blue and gold and if we reinstalled the huge, gaudy cult figure of Athena, festooned in bracelets, rings and necklaces, we could not avoid the question that threatens our whole concept of the classical: did the Greeks have bad taste?
>
> (Horne 1984: 29)

The Polynesian Cultural Centre, the most popular tourist attraction in Hawaii, builds its exhibits around various cultural groups of the South Pacific and their traditional dwellings, dances and crafts. There is some attention to accuracy in terms of the cultures presented, but the performers –

college students from the Mormon College at Laie – may or may not be from the group they are depicting. Fijians and Tongans may be doing Samoan dances. The individuals are merely acting (Brameld 1977). Does it matter? In this instance, this writer would argue 'staged authenticity' (MacCannell 1979) may be preferable to making people's actual lives a tourist attraction.

The more interesting question is 'who gets what' from this centre? Are the performers properly paid for their labour? Who controls what is presented? Do the cultures presented accurately portray current life in the South Pacific or do they contribute to what Dann refers to as 'museumization, the freezing of a heritage and the selling of the frozen product' (Dann 1994: 59). Do the visitors get an appreciation of South Pacific cultures or a sense of superiority over these societies? We do not know.

Some groups are adept at seizing the tourism initiative by creating an affordable, tourist-friendly heritage. The Ute of Southwestern Colorado is one group that has taken control of its evolving heritage and its tourist future in ways scarcely traditional. Up until modern times the Utes were well-known for their beadwork, but contemporary Utes found beadwork tedious and unrewarding. What to do? They decided to learn pottery from a non-Indian, determine a few designs and colours that would define Ute pottery and went to work. In 1986 when this writer visited their factory cum tourist shop they had just begun. Today one finds 'traditional Ute pottery' featured in specialty catalogues. It is authentic but the tradition is only twelve years old.

The Toraja of Indonesia are another group that have absorbed those useful components of their tourist image and made them their own. That has engendered some conflict among those neighbour groups who wish to share the economic windfall from tourism (Adams 1995: 143–53).

The political power of various groups may well control not only whose interpretation and definition of authenticity prevails but also what will be saved or remembered at all.

MISERY LOVES COMPANY

'a painful past is good box office'.

(Horn 1997: 54)

There is no question that heritage tourism today in many societies is much more willing to confront the shameful legacies of the past and the inability of government policies to live up to the ideals immortalized in more traditional heritage sites. What is more problematic is whether the messages are primarily motivated by greed, 'milking the macabre' as Dann puts it (1994), or the need to re-examine the tragedies so tourists can learn from them how not to behave. Equally hard to gauge has been the message received by the tourists. The labels 'dark tourism' or 'thanatourism' have been put on such tourism, but there have been relatively few sites assessed for their impact (Seaton 1996 and 1997).

One that has is Dachau Concentration Camp with its shrine and museum. From 1965 to 1988 more than thirteen million visited the scene of the extermination of millions of Jews, gypsies, and others the Nazis deemed undesirable. Sixty per cent of the visitors were international visitors; many of the others were German school children. No one surveyed their motivation in coming, their impressions of the camp nor did any follow up of the impact of these visits. Their motives have been assumed to be (1) personal, perhaps seeking the names of relatives, (2) political, since political parties persecuted by the Nazis often hold meetings there, (3) historical and educational, as in the case of the many school children brought there and (4) humanitarian, as it is the site of Jewish–Christian groups and youth camps.

What *is* known about Dachau's impact is that the residents of the city are becoming increasingly unhappy with being associated with genocide. Yet, thirteen million visitors do generate impressive economic benefits. In an effort to keep the tourists coming but to blur the message, there have been specific efforts made to put more emphasis on the local resistance to Nazism and the activities of the anti-fascists from 1918 to 1945 (Hartmann 1989: 41–7).

The Gestapo Prison Museum in Berlin, the Japanese Prisoner of War Museum in Singapore, Auschwitz in Poland, the Inquisition Museum in Lima, the KGB museum in Moscow and even the Tower of London are other examples of this type of heritage tourism.

In this regard two other museums need to be mentioned which have opened in the 1990s: the Holocaust Museum in Washington DC and the Beit Hashoah Museum of Tolerance in Los Angeles. Both are designed to leave a lasting impression of the evil of bigotry by the various ways in which they encourage visitors to identify with the victims of persecution. Reactions differ as to the probable lessons such museums convey (Rosenthal 1993: 4–5A). Horne, who studied similar museums in Europe, is not so sanguine about their impact:

> In monuments to the victims of fascism, what is novel are reminders of the concentration camps. Museums display . . . photographs of incinerators, of patient queues obligingly waiting to be hanged, shot or gassed, and the piles of the dead . . . With a clear eye one can see the concentration camp museums not as monuments to the heroic, but to the sufferings of passivity: a record of the triumph in power of one of the possible forms of a modern industrial state. To the tourists on a horror pilgrimage, however, this can seem exceptional – something that ended with the Nazis.
>
> (Horne 1984: 244)

The Museum of Tolerance avoids Horne's critique by stressing the point philosopher Hannah Arendt made about the 'banality of evil'. It includes the Holocaust in Germany but also similar bloodbaths in Cambodia, Rwanda, and elsewhere. The very unexceptional quality of hate groups throughout the world attempts to send home the message that intolerance is a virus to which no society is immune.

Professor Mary Philips' reaction illustrates a quite different opinion from Horne's about the messages received from such museums, and both impressions seem to beg for more empirical research. Philips writes (1994: 43) that as she left the Museum of Tolerance:

> I saw a sign whose message I hope is burned in my memory. It said, 'As long as we remember, there is hope . . .' There is not only the danger that some will forget what happened. There is another danger that some people, for whatever reason, never truly understood what was happening all around them . . . It is essential that educational experiences such as this . . . be provided for future generations, for those who weren't connected to the times in which we lived.

Clearly, Philips got the message being sent. Authorities are also using the Museum of Tolerance as a place to send juveniles involved in hate crimes. The director of the Museum felt that the visit had a positive but sobering impact on the individuals sent (Personal interview March 1994).

In Kuala Lumpur, Malaysia, at least one new museum, the Pudu prison, is being used to instruct children about the evils of drugs and the discomforts of prison life. Mock executions and real videos of canings drive home the message (Personal tour 1997).

Groups need to know if such places have the intended impact before they proliferate further, sometimes with public funding. In a violence-satiated environment of horror films and television carnage one could argue that it is the very generation the museums are trying to reach most that may be least susceptible to its values. The line between what is real and what is not is increasingly blurred to a generation attuned to a diet of mayhem and virtual reality.

Nor need we assume that only youths will be attracted for the wrong reasons to such grim tourism. While the phenomenon of tragedy tourism has accelerated recently, in its tackier forms it has been around for some time: 'In Paris at the turn of the century, sightseers were given tours of the sewers, the morgue, a slaughterhouse . . .' (MacCannell 1979: 57).

Political will to make such heritage attractions accessible to the public may vary from society to society and according to who are cast as the villains in the story. Given the penchant for privatization around the globe, we may expect more and more heritage sites to reflect private aspirations rather than national or other public considerations. Is this consistent with diversity or the balkanization of national identity? One may take encouragement from the Mashantucket Pequot Tribe in Connecticut who seized the initiative and raised the money for a $135 million museum to commemorate the 1637 massacre of tribe members (Horn 1997: 56). Would we be as sanguine with various anti-government militias running museums extolling white supremacy, secessionist and conspiracy agendas? Perhaps they already are.

While it may be more politically palatable if the horrible events remembered are the work of other countries or groups, public institutions are belatedly acknowledging painful episodes in their own history. '*Mea culpa*' tourism as it might be called is perhaps cathartic for the dominant group and some solace to others. The US National Park Service has recently added the Manzanar National Historic Site to the park system. It commemorates the thousands of Japanese-Americans who during World War II were uprooted from their homes, farms and businesses solely on the basis of their Japanese ancestry and interned for years in guarded camps. The Wounded Knee Battlefield is also an American historical site acknowledging the US cavalry's slaughter of 300 Sioux. Similarly, the Trail of Tears stretching from Tennessee to Oklahoma marks the uprooting of 15,000 Cherokee, many of whom died en route.

CONCLUSIONS: HERITAGE TOURISM AND THE POLITICS OF POWER

Several broad issues have been explored in this chapter. First, not only is the balance of power shifting somewhat from dominant groups to the formerly marginalized, but other changes form an increasingly decentralized, privatized, and yet often globalized policymaking environment. As the heritage sites have changed, so have the opportunities for two opposing tourist experiences. The visitors now have an unprecedented opportunity to interact with and make an impact on the sites at many heritage destinations. But at the same time, the place itself has an increased ability to control and manipulate the tourist's visit.

Second, there has been a virtual explosion in the number of heritage sites that commemorate the powerless and the overlooked. Some of these places remember tragic episodes, others the striking achievements of groups long ignored. Third, as the number of groups represented in heritage sites has grown, there has been an intense struggle over the content of that representation. It has revolved around issues of accuracy, emphasis, authenticity and political message.

Fourth, the final trend examined is the growing popularity of tourism focused on sad or tragic experiences. Interpreting such sites as vehicles for increasing understanding and reducing racism and bigotry is a comforting but neglected hypothesis that needs to be pursued. Such places could be powerful, just as television and movies have been charged, in producing 'copycat crimes' and entertaining with tales of the lurid and grotesque.

Each issue raised can be examined according to disciplinary emphasis but several central political questions lead back to 'Who Gets What, When and How' (Lasswell 1936).

> Though a declared aim of all of them [all kinds of museums] was educational, one of the unintended functions of these museums was to put most people in their place. A much quoted survey carried out in France in 1966, suggested that eight out of ten working people who visited an art museum associated it with a church; two-thirds could not remember the name of a single work.
>
> (Horne 1984: 16)

In the case of some exhibits, forgetting might be the best one might hope for. The Smithsonian Museum of American History had a provocative exhibit in the late 1980s on World Fairs. At one such fair shortly after the American conquest of the Philippines in 1898, tribal people were brought to the US as an exhibit. Used to the tropics, several died of exposure. Grotesque as such an exhibit might seem today, the 'zooification' of indigenous peoples goes on around the globe. Such exhibits are thought by many to justify imperial designs and pacification campaigns or simply make the tourist feel prosperous by comparison:

> The world's fairs of this era preached that white men's manliness fueled the civilizing imperial mission and in turn, that pursuing the imperial mission revitalized the nation's masculinity. White women were meant to come away from the fairs feeling grateful . . . without the Samoan, Filipino and other colonized women, neither male nor female fair-goers would have been able to feel so confident about their own places.
>
> (Enloe 1989: 27)

Today's museums, it might be argued, attempt more thematic exhibits, but since museums and exhibits increasingly appear to be more ideological in content, it would be worth studying what the tourist receives from the experience. When Philippine President Corazon Aquino made a museum of Malacanang Palace where her corrupt predecessor, Ferdinand Marcos, had lived, the expectation was that the impoverished public would be outraged at the lavish excesses and her own seizure of power legitimized. Some got the intended message, but many others were simply dazzled by the fantasy life of the dictator. The palace tours ceased (Richter 1989).

Nor does the current era seem overcome with a consideration for the very different memories of fellow citizens. The 500th anniversary of Columbus'

'Discovery of America' was seen by the US government as a great tourism promotion opportunity. What was neglected was the fact that many African-Americans, and most Native Americans and many in the Caribbean regret and mourn the so-called 'discovery'. The aboriginals of Australia also did not celebrate the bi-centennial of the first ships arriving with the white British prisoners who would colonize that country (Altman 1989: 456–76).

Case studies are needed to probe who gains and loses from heritage tourism. There are a lot of assumptions about it with very little empirical data. We do know that a double standard still persists in what is legitimate to investigate and make into attractions. Excavations and tourist attractions based on burial or sacred indigenous sites are common. Exhuming pioneers and pilgrims for tourist sites would be unthinkable!

Ultimately, the most important political issue surrounding heritage tourism is whether even an extensive exposure to it leads to a better informed or perhaps more tolerant individual. If so, schools could better make the case for excursions and parents might be more ready to include some heritage sites in their travel. Public and private funds for such places might have more support. Studies have shown again and again that education followed by income are the best predictors of tolerance toward others. In our increasingly diverse societies does heritage tourism have a role to play in providing that education? Can episodic, out-of-context, historical exhibits yet create a foundation to inspire future learning? Those who can say what impact such tourism has will be needed to inform public policy and private development decisions for the future of heritage tourism.

References

Adams, M. (1995) 'Making up the Toraja? The appropriation of tourism, anthropology, and museums for politics in Upland Sulawesi, Indonesia', *Ethnology* 34 (2) Spring: 143–53.

Altman, J. (1989) 'Tourism dilemmas for aboriginal Australians', *Annals of Tourism Research* 16 (4): 456–76.

Ashworth, G.J. and Tunbridge, J.E. (1990) *The Tourist-Historic City*, Belhaven, London.

Boyer, M. (1992) 'Cities for sale', pp. 181–204 in Michael Sorking (ed.) *Variations on a Theme Park*, New York: Hill and Wang.

Brameld, T.B. (1977) *Tourism as Cultural Learning*, Washington DC: University Press of America.

Bruner, E. (1993) 'Lincoln's New Salem as a contested site', *Museum Anthropology* 17 (3): 14–25.

Dann, M.S. (1994) 'Tourism: the nostalgia industry of the future', in William F. Theobald (ed.) *Global Tourism: The Next Decade*, London: Butterworth–Heinemann Ltd.

Duara, P. (1997) a review of Jonathan Boyarin (ed.) *Remapping Memory: The Politics of Timespace*, Minnesota: University of Minnesota Press, in *Journal of Asian Studies* February, 56: 141–2.

Edgell, D. Sr (1993) *World Tourism at the Millennium*, (April) Washington DC: US Dept of Commerce.

Eisenstodt, J. (1995) 'Should planners have a social conscience', *Meeting News* April 24: 15.

Enloe, C. (1989) *Bananas, Beaches, and Bases*, Berkeley: University of California Press.

Fredericks, A. (1992) 'Thinking about women in travel', *Travel Weekly* July 2: 12.

Garnet, S. (1994) 'Buffalo soldiers: rugged men of the west', *Dollars and Sense*, January: 56–63.

Graburn, H.H. (1995) 'Tourism, modernity and nostalgia', pp. 154–77 in A. Ahmed and C. Shire (eds) *The Futures of Anthropology*, London: Athlone Press.

Hartmann, R. (1989) 'Dachau revisited: tourism to the memorial site and museum of the former concentration camp', *Tourism, Recreation, Research* 14 (1): 41–7.

Hoivik, T. and Heiberg, T. (1980) 'Centre–periphery tourism and self-reliance', *International Social Science Journal* 1: 69–98.

Holloway, N. (1995) 'Museum peace', *Far Eastern Economic Review*, February 2: 32–3.

Horn, R. (1997) 'Through a glass darkly', *US News and World Report*, May 26, 122 (20): 54–60.

Horne, D. (1984) *The Great Museum: The Re-Presentation of History*, London: Pluto Press.

Ireland, M. (1993) 'Gender and class relations in tourism employment', 20 (4): 666–84.

Lasswell, H. (1936) *Politics: Who Gets What, When and How*, New York: Whittlesby House, McGraw Hill

'Left at the wall' (1992) *The Economist*, October 31.

Leo, J. (1994) 'The national museum of pc', *US News and World Report* October 10: 21.

'Lifting the veil on Arab women artists' (1994) *Women in the Arts*, XI (4): 1–4.

MacCannell, D. (1979) 'Staged authenticity: arrangements of social space in tourist settings', *American Journal of Sociology* 79: 589–603.

'Maya Lin: A strong clear vision' (1996) Public Broadcasting System, November 26.

Murphy, P.E. (1992) 'Urban tourism and visitor behavior', *American Behavioral Scientist* 36 (2): 200–11.

'Never again' (1993) *The Economist*, May 1.

Nietzsche, F. (1988) *On the Advantage and Disadvantage of History for Life*, Hackett Publishing Co, 5th Printing, Indiana.

Norkunas, M.K. (1993) *The Politics of Public Memory – Tourism, History and Ethnicity in Monterey, CA*, New York: State University of New York.

Philips, M. (1994) 'The Beit Hashoah museum of tolerance', *National Forum* LXXIV (1): 31–3.

Richter, L.K. (1989) *The Politics of Tourism in Asia*, Honolulu: University Press of Hawaii.

—— (1994) 'Exploring the political roles of gender in tourism research', in William Theobald (ed.) *Global Tourism: The Next Decade*, New York: Butterworth.

—— and Richter, L. (1995) 'Reinventing government abroad', a paper presented at the American Society for Public Administration meeting in San Antonio, Texas.

Rosenthal, F. (1993) 'New holocaust museum offers tour through hell', *Wichita Eagle* April 19: 4–5A.

Seaton, A.V. (1996) 'Guided by the dark: from thanatopsis to thanatourism', *International Journal of Heritage Studies* 2 (4): 234–44.

—— (1997) 'War and tourism: the paradigm colussus of Waterloo 1815–1914', paper presented at the War, Terrorism and Tourism Conference in Dubrovnik, 27 September, 1997.

Wagner-Pacifici, R. and B. Schwartz (1991) 'The Vietnam veteran's war memorial: commemorating a difficult past', *American Journal of Sociology* 97 (2) September: 376–420.

Washburn, W.E. (1995) 'The Smithsonian and the Enola Gay', *The National Interest* Summer: 40–9.

Wills, G. (1997) 'Carved in stone', *The Washington Post National Weekly Edition* May 5: 21.

WTO (1998) 'Tourism growth slows due to Asian financial crisis', WTO News, March–April: 1–2.

Urry, J. (1992) 'The tourist gaze "revisited"', *American Behavioral Science* 36 (2) November: 172–86.

Other sources:

Personal interviews in Canberra, Australia and in Manila, Philippines (1987 and 1992) and Los Angeles, US (1994).

Personal tours of Pudu Prison (Kuala Lumpur), Museum of Tolerance (Los Angeles), Australian War Museum (Canberra), Malacanang Palace (Manila), Smithsonian Museum of American History, Holocaust Museum, and National Museum for Women and the Arts (Washington DC).

8

THE SPREAD OF CASINOS AND THEIR ROLE IN TOURISM DEVELOPMENT

William R. Eadington

INTRODUCTION

Casino gaming industries in many countries have experienced substantial growth and expansion in the 1990s. Much of this has been a direct result of explicit strategies adopted by state, provincial or national governments which believe that casinos can be an important catalyst in creating or otherwise stimulating growth and tourism within their borders. However, the success of using casinos as a growth strategy and a tourism strategy has been mixed. The actual impacts of casino developments and their contributions to tourism objectives have depended on a variety of circumstances.

The form of legal gambling that is most associated with tourism is casino gaming. Other popular forms of gambling – such as lotteries, wagering on racing, charitable gambling, and non-casino located gaming devices – cater predominantly to local markets and therefore have little direct impact on tourism or tourism development. On the other hand, famous historic casino centres, such as Las Vegas, Monte Carlo, Sun City and Macao, have attracted visitors from neighbouring or distant states or countries as their main source of business. Indeed, Las Vegas, which in 1997 attracted over thirty million visitors per year to its 105,000 hotel rooms and myriad casino and entertainment facilities, had become an ideal tourism destination resort, centred around casinos.

The economic ramifications of the recent spread of casinos and legal gambling have been discussed elsewhere by this author (e.g. Eadington 1995, 1998a; Eadington and Cornelius 1997). Inspired by the phenomenal growth and economic success of Las Vegas, many other jurisdictions have authorized numerous forms of legal gambling – including casinos – for various reasons, including tourism development.

This analysis examines the wide variety of casino ownership and regulatory regimes that have been introduced – especially since the mid-1980s, in light of their promises and their ability to contribute toward strategic tourism objectives. The first section looks at the recent experiences in the United States, Australia and Canada in the approaches they have taken in legalizing casinos. The next section discusses economic and social impact issues that are implied by the ways casinos have been implemented, especially in light of the political and economic tensions casinos create under varying circumstances. Common social and economic impacts that are associated with casino developments are examined, and factors that lead to increased or diminished controversy with regard to casinos are scrutinized. This section also looks at competitive relationships that exist among various types of gaming alternatives that are considered 'hard gambling'. The distinction between 'hard gambling' and 'soft gambling' is made by the British Gaming Board and the Home Office as a justification for stricter controls placed upon those forms of gambling at which an individual can risk and lose a significant amount of money in a short period of time (i.e. casinos, slot machines) as opposed to those which are not as financially dangerous (i.e. lotteries, bingos, raffles, 'amusement with prize' machines) (*Report of the Gaming Board for Great Britain 1996/97*). The final section looks at some possible futures that casino gaming might encounter in its various manifestations throughout the world.

THE RECENT PROLIFERATION OF LEGAL CASINO GAMING

When gambling is taken from illegal to legal status, it taps into a strong latent demand among some segments of the population who demonstrate a willingness to spend considerable time and money on the activity. It is apparent that legal permitted gambling has much broader appeal than gambling that takes place only in illegal settings. As a result, jurisdictions that have legalized casino-style gambling have seen new gaming industries emerge having substantial impacts on consumer spending patterns, as well as on competing and complementary industries. In varying degrees, these impacts can be noted in the examples noted below.

The United States

In the United States, casino gaming was prohibited everywhere but Nevada from 1931 to 1978, and was then authorized only in Nevada and New Jersey between 1978 and 1989. By the late 1990s, however, the number of states offering casinos expanded to more than twenty-five. Numerous types of casinos became legal, including casinos on riverboats, in historic mining towns, in urban and suburban locations, and on Indian lands.

The size of all commercial gaming industries in the United States, as measured by *gross gaming revenues* – total revenues after payment of prizes, the same as aggregate customer losses or expenditures – grew from $10 billion in 1982 to $47 billion in 1996 (Christiansen 1997). The casino industry generated gross gaming revenues of nearly $25 billion in 1996, up from $6 billion only ten years earlier.

There are wide variations in the ways in which casino gaming is offered in the US. True 'full service' destination resort casino hotel entertainment complexes can be found only in Nevada and Atlantic City, New Jersey. Indeed, by the end of the 1990s, the standards for such resorts in Las Vegas call for facilities with 3,000 or more hotel rooms; unique, often spectacular, architecture; extensive entertainment offerings; indoor and outdoor recreational options; extensive culinary and shopping experiences; and, of course, 'state of the art' gaming opportunities. The price tag for building a new casino property for the Las Vegas Strip tourist market passed $500 million in the 1990s, with some new properties flirting with capital outlays approaching $2 billion. This trend toward bigness, entertainment and diversity is being mimicked in other casino centres such as Atlantic City, Reno, and Biloxi and Tunica County, Mississippi, but at not nearly the same scale.

The destination resort casino industry in Las Vegas is the largest and most dynamic of any casino industry in the world. Mega-casino complexes have changed the face and image of Las Vegas from *gambling* to *casino entertainment* in the 1990s. The new casinos along the Las Vegas Strip that opened between 1989 and 1997 – The Mirage, Excalibur, Treasure Island, MGM Grand, Luxor, Monte Carlo, and New York New York – accelerated Las Vegas along an unprecedented growth path and transition. By 1997, nearly 50 per cent of revenues accruing to the largest nineteen casinos along the Las Vegas Strip came from non-gaming sources. The popularity of this approach is reflected by the fact that in 1997 over 80 per cent of all the profits earned by the more than 200 major casinos in Nevada came from this clustering of mega-casino complexes (State Gaming Control Board 1998).

But the full service competitive casino–hotel complex model developed first in Nevada and, modified by Atlantic City, did not gain favour in other US jurisdictions. Rather, various hybrids were introduced in some parts of the country, and then copied by other states. Small-stakes mining town casinos were authorized by referendum in Deadwood, South Dakota in 1988, which was followed by a similar successful statewide initiative in Colorado benefiting the former mining towns of Central City, Cripple Creek, and Blackhawk in 1990 (*Economic Development Review* 1995; *Journal of Travel Research* 1996; Kent-Lemon 1996; Meyer-Arendt and Hartmann 1998). The mining town casinos are characterized by size limitations, restrictions on the number of gaming devices or gambling tables per location, and by requirements that casinos be permitted only in historic buildings in established commercial areas. Furthermore, a five dollar maximum wager requirement precludes the profitable development of table games, so casinos in these towns are heavily dependent on slot machine revenues.

As a result of such statutory limitations, opportunities for growth and evolution of mining town casino industries have been stifled. In 1997, Deadwood, South Dakota's eighty or so casinos produced gaming revenues of $43.5 million, and the dozens of casinos in Colorado's towns generated gaming revenues of $430.6 million. In contrast, two Indian casinos in rural Connecticut – Foxwood's and the Mohegan Sun – experienced slot machine winnings in the amount of $943.8 million, and total gaming winnings of nearly $1.5 billion (Salomon Smith Barney 1998).

Riverboats provided the milieu for casinos authorized by legislation in Iowa, Illinois, Louisiana, Missouri and Indiana between 1989 and 1993. The fundamental rationale was that by permitting casino gaming only on boats, and – initially – only when boats were sailing, social damages that might accrue from the presence of casino gaming in communities in these states would be mitigated. In reality, the effect of allowing casino gaming only on riverboats has inconvenienced casino gaming customers and increased the costs of operations to casino owners. Indeed, there is virtually no desire on the part of riverboat casino customers or operators to sail.

A variant of riverboat casinos was introduced in Mississippi in 1992. However, unlike other states that authorized riverboat casinos, Mississippi's boats were permitted from the outset to remain dockside. Soon thereafter, the law was more liberally interpreted to define a 'boat' for gaming purposes as any structure built over the water along permitted waterways. In other respects, Mississippi statutes and regulations parallel those of Nevada. As a result, casino regions in Mississippi – especially in Tunica County near Memphis, and Biloxi on the Gulf Coast – have experienced considerable evolution toward true tourist-oriented destination resort enclaves. Gross gaming revenues in Mississippi grew rapidly from the opening of the first casinos in 1992, climbing to just under $2 billion in 1997.

Urban casinos have had a rocky time in the United States in the 1990s. In New Orleans, an exclusive franchise monopoly casino that opened in May 1995 declared bankruptcy by November of the same year. By 1998, the owners were still trying to reach an accord with the city and state that would allow them to reorganize and reopen. The causes for the New Orleans bankruptcy were many, but perhaps most important were the substantial tax obligations – annual guarantees of $100 million to the state – placed on the operation by statute, coupled with prohibitions against the casino offering hotel accommodation, restaurants or retail outlets on premises. Though it was initially hoped that an urban casino in New Orleans would broaden the city's tourist and convention appeal, the political failures and poor performance of landbased and riverboat casinos in New Orleans brought into question the economic viability of casinos when they must compete with many other existing tourist activities. Riverboats in New Orleans have also not fared well. As of mid-1998, there were no operating riverboats near the centre of the city (though there were some successful suburban riverboat casinos in operation), and a number of riverboat casinos had either ceased operations, cancelled construction, or declared bankruptcy.

The city of Detroit began the process of introducing urban casinos following a favourable statewide referendum in Michigan 1996. The referendum allowed for three casinos to be developed within the city of Detroit. The main objectives were to stimulate tourism and convention business to the city, to create jobs for city residents and tax revenues for local governments, and to stem the flood of US dollars that were crossing the international border to the provincially owned casinos in neighbouring Windsor, Ontario, Canada. Visitors to the Windsor casinos are mainly residents of the greater Detroit metropolitan area and, as such, are not really tourists. The challenge for Detroit – a somewhat distressed industrial city with a legacy of high crime rates and declining neighbourhoods – to develop as an actual tourism draw based on casinos will be formidable.

Indian gaming is the other major form of casino-style gaming in the United States, with over 125 casino-style operations in over twenty states in 1997 (US General Accounting Office 1997). Indian gaming came into existence as a result of a 1987 Supreme Court decision and the 1988 passage of the Indian Gaming Regulatory Act (Eadington 1990; Kelly 1993–94). The general form of Indian casinos – in terms of size, location, permitted games and other significant circumstances – is highly dependent on the state in which the tribal lands are located. Negotiations between tribes and state officials leading to compacts, various court interpretations and proximity to population centres have had considerable influence on the resulting Indian casinos. For example, Foxwood's Casino in Connecticut is the largest and most profitable casino in the world, with gross gaming revenues approaching $1 billion. It holds that status because of a shared monopoly (with another Indian casino, the Mohegan Sun) for casino-style gaming in New England, and it is relatively close to the major metropolitan concentrations around the cities of Boston and New York. Its compact with the state permits all forms of casino games permitted in Nevada or Atlantic City, along with a full array of gaming devices. Furthermore, there are no other legal forms of casino-style gambling – such as video lottery terminals or video poker machines inside or outside casinos – permitted in the region.

For the most part, Indian casinos service customers within regional markets. A significant portion of customers – well over 90 percent – to Indian casinos reside within 100 miles of the casinos. However, to the extent that Indian tribes are politically and socially independent from the states in which they reside, Indian casino customers are effectively 'tourists', and tribal casinos are major sources of income and wealth for their respective tribes.

Variations in casino-style gaming have evolved in a number of other states as well, and these tend to have virtually no tourist component among their customer bases. Slot machines, video poker machines and other types of electronic gaming devices are permitted in certain non-casino locations in the states of Montana, Louisiana, South Carolina, South Dakota, Nevada and Oregon. These are sometimes referred to as 'slot route operations', sometimes as 'video lottery terminals'. Gaming devices are usually permitted only in age-restricted locations such as bars and taverns, though Nevada permits slot machines and video poker machines in a wide variety of businesses, including restaurants, supermarkets, drug stores and car washes. Louisiana permits video poker machines at truck stops, race tracks and off-track betting parlours. The most important implication of such permitted gaming as far as tourism is concerned is to cut into the market potential for more tourist-oriented destination resort casinos.

In recent years, race tracks in the states of Iowa, Delaware and New Mexico have been successful in persuading legislatures to allow them to operate large numbers of slot machines. When this is permitted, the net effect has been to transform the tracks from racing establishments to casinos that happen to run races. At Prairie Meadows in Iowa, for example, over 90 per cent of gaming revenues in 1997 – $127 million – came from slot machines. In similar fashion, the states of West Virginia and Rhode Island permit video lottery terminals at the tracks. The Hollywood Park race track in California operates a card club casino at the facility. In light of the inherent popularity of casino-style gambling, along with the recent financial distress of the racing industry, it is expected that political pressure for more casino-style gaming at race tracks will continue. As with urban casinos, Indian casinos and gaming devices located outside casinos, slot machines at tracks cater more to a local and regional customer market than to a true 'tourist' market.

Australia

Other countries have had similar experiences as the United States with regard to the expansion of permitted gaming. Australia authorized its first legal casino in the state of Tasmania in 1972. Over the next decade, three additional casinos were introduced in Tasmania and the Northern Territory, in remote and sparsely populated parts of Australia. The continuing prohibitions against casinos in the major cities and states ensured tourism travel to these casinos. However, Australia became the first country to permit American style casinos in urban locations when it legalized monopoly casinos in Perth, Adelaide and the Gold Coast near Brisbane in the mid-1980s.

By 1997, casinos had opened in every major city in the country, as well as in some established tourism centres, such as the Gold Coast and along the Great Barrier Reef. Australian casinos are, for the most part, exclusive franchise monopolies run by private sector companies located in or close to city centres. It was little surprise that the tourism draw of the original remote 'destination resort' casinos began to decline when new urban casinos opened their doors.

Most states in Australia have also authorized widespread placement of electronic gaming devices in hotels, bars and taverns. In New South Wales, slot machines – locally known as *pokies* – have operated legally in not-for-profit arcades and clubs since the 1950s. By the 1990s, widespread placement of such devices, along with video poker machines, had taken place in every state except for Western Australia.

Though one of the major selling points of the urban casinos in Australia was the potential attraction of such facilities for international tourism, the major customer base for the casinos is the local metropolitan market. Exceptions to this are the casinos in tourist regions such as the Gold Coast and the Great Barrier Reef, and some casinos have been very successful in attracting premium players from South-East Asian capitals such as Hong Kong, Singapore, Taiwan, Bangkok, and Kuala Lumpur. Though few in number, such visitors wager extremely high amounts, especially at the game of baccarat, and they therefore can have a significant impact on casino earnings and foreign exchange.

Canada

Casino gaming came to Canada by a slightly different path than it did to the United States or Australia. Canadian provinces had established lotteries throughout the country as a revenue raising mechanism in the 1970s, but did not get into the casino business in any serious way until the 1990s. There did exist in the western provinces a so-called 'charitable casino industry' from the 1970s onward, where small-stakes table games temporary casinos would be used to raise funds for charity and not-for-profit organizations (Campbell and Lowman 1990; Campbell 1994). Government-owned monopoly casinos were introduced in various urban and destination resort areas – including Montreal, Halifax, Winnipeg, Hull, Niagara Falls, Regina and Windsor – between 1990 and 1997. For the most part, Canadian casinos follow an urban monopoly casino model similar to Australia's, though ownership is by government rather than the private sector.

The motivation for some Canadian casinos, such as Windsor, Niagara Falls and Hull, was to capture out-of-province or foreign customers by locating the casino adjacent to major population centres in the United States or at a provincial border. In other cases, such as Montreal and Halifax, the intent was to provide an additional amenity to simulate tourism, but the effect was to create casinos primarily frequented by local citizens.

In one respect, the Windsor and Niagara Falls casinos might be classified as 'tourist' casinos, as most of their customers are Americans. However, these are 'border casinos' that are successful because of prohibitions against casino gambling on the other side of the border, and virtually all of the 'tourists' in the casinos are day-trip visitors attracted by the opportunity to gamble. Not surprisingly, the City of Detroit – across from Windsor – legalized three casinos through a statewide referendum in 1996, though the permanent casinos will probably not open until after the year 2000.

By 1997, a number of Canadian provinces had also introduced video lottery terminals widely distributed in cities and communities. However, the customers for this form of gambling are almost all 'locals', so there is little direct linkage to tourism as a result.

REASONS FOR THE SPREAD OF CASINOS

The spread of casinos in the United States and in other countries occurred for a number of reasons. First, lotteries had generally preceded permitted casinos and whetted government appetites by demonstrating the popularity and revenue-generating potential of gambling. Lotteries also softened the public's attitude regarding the risks associated with gambling.

The major arguments against permitting 'hard' gambling – such as casinos – centre around three types of argument: links between gambling and criminal influences; claims of the immorality of gambling; and social consequences of problem and pathological gambling. The perception of linkages between casinos and crime has diminished in many countries over the past couple of decades as casino companies have become more mainstream and respectable. Furthermore, regulatory bodies have become more effective in fulfilling public policy mandates and in protecting a variety of public concerns.

Morality arguments against gambling have diminished in many jurisdictions, perhaps reflecting the weakened influence of organized religion in many societies. On the other hand, the issue of problem or pathological gambling remains as the most significant challenge which confronts permitted gambling, especially 'hard gambling'.

It is often the economic dimension that has been the driving impetus behind permitting most forms of gambling, especially casinos. However, the desired outcomes of economic development and tourism stimulation have not been universal. In order for significant economic stimulation to occur, a large proportion of custom must come from outside the region where the casinos are located. Alternatively, casino facilities that cater primarily to locals will not have a substantial impact on growth unless they heavily draw business from local residents who would otherwise leave the region in order to gamble.

In general, most customers of urban casinos have not been tourists. Many new casino jurisdictions in the United States provide 'casinos of convenience' which cater predominantly to residents of the area where the casinos are located. In such cases, there has been little net economic stimulation to the area, though the casinos themselves have been substantial revenue generators. The same pattern is observable with casinos in most other countries that have recently authorized casinos.

The long-term future of casinos and other forms of 'hard gambling' depends to a great extent on society's acceptance of gambling as a legitimate consumer pursuit. The last half of the twentieth century has been characterized by a steady increase in the degree of acceptance of the general public in the activity of gambling, in spite of the fact that the other vices – alcohol, tobacco, illicit drugs – have been under increasing criticism and sanctions in much of the world.

Permitted gambling has been embraced more for the ancillary economic benefits it creates than for the customer demands that it fulfills. This perspective might be challenged in the future, however, as gambling becomes more localized and less tourist-oriented. Furthermore, new technologies will increasingly bring gambling into the home – through the Internet, through interactive television wagering systems – whether or not such activities are legally sanctioned. This could further erode gambling's role in tourism and increase public concerns about the wisdom of permitted gambling (Eadington 1998b).

Thus, if permitted gambling continues to expand in society, then the role of gaming in tourism will likely decline unless gambling, especially casino-style, becomes part of a wider range of complementary entertainment offerings. This is a formula that has been well developed by Las Vegas, but not in many other locales. Usually, casinos become tourism generators primarily because of prohibitions of gambling in places where people live. As those prohibitions disappear, then much of tourism-based gambling will diminish as well.

The legislative explosion permitting casino gaming from the mid-1980s onward was a result of a variety of factors. First, many societies have cultivated more positive attitudes about the acceptability of gambling in general. This is reflected in the United States by public attitude surveys that indicate a high proportion of Americans who believe that casino gambling is an appropriate form of entertainment for themselves, or if not for themselves, it is still appropriate for others. Second, governments and policy makers have been willing to utilize this controversial industry as a means to achieve broader ends. Third, there has been a substantial broadening of the base of customers for whom casino gaming has become an entertainment activity of choice. In the United States, over 142 million casino visits were made in 1996 (Harrah's Entertainment Inc. 1997). Fourth, there has been a decline in the perceived and real influence of organized crime and other notorious elements in existing casino industries in the United States and abroad and a corporatization of the casino industry (Johnston 1992).

WILLIAM R. EADINGTON

LEGAL AND REGULATORY STRUCTURES AND SOCIAL IMPACTS

The spread of casinos in various countries, states and provinces has occurred with a diverse mix of strategic approaches. In effect, many such authorities have come to the position that permitting gambling is preferable to prohibition, but having reached that position, they still retain reservations regarding the potential adverse impacts casinos might create for society. Thus jurisdictions have undertaken fundamentally different approaches in introducing casino and casino-style gambling on such issues as ownership, market structure, permitted locations and operating constraints. These different approaches reflect diverse philosophic, political and cultural views on how best to exploit the gains associated with allowing casinos, while at the same time mitigating negative side effects and political backlash related to permitted gambling.

The types of negative social impacts that have raised the greatest concerns have been linkages between casinos and casino-style gambling and organized crime; neighbourhood crime and other crimes against property, such as burglaries, break-ins and embezzlements; and family-related crime and disruptions, such as child abuse, spouse abuse, suicides and divorce. One of the fundamental realities regarding economic and social impacts associated with gambling is that economic impacts – which tend to be positive – are quantifiable, tangible and measurable; whereas social impacts – which tend to be negative – are qualitative, elusive and very difficult to measure. Thus, one can often readily account for positive economic impacts within new casino industries, such as visitations, revenues, tax collections, jobs created and new investments undertaken. However, it is very difficult, if not impossible, to come up with meaningful measures of the incidence of many social impacts attributable to an expanded presence of permitted gambling.

The above point notwithstanding, various claims have been made in recent years, primarily by opponents of legalized gambling, about the incidence of negative social impacts linked to casinos and casino-style gambling, and the costs they create for society at large (Goodman 1995; Grinols and Omorov 1997; Thompson, Gazel and Rickman 1997; Kindt 1998). Such claims have been based on questionable research methods, unverifiable and untraceable numbers, selective data interpretation, incorrect concepts, or just shoddy analysis. Nonetheless, claims made in the popular media tend to take on a life of their own, especially in the politically charged environment that has surrounded recent debate on the overall effects of permitted gambling on society (Kelly 1997; Walker and Barnett 1997).

Measuring social impacts associated with casino gaming is inherently difficult. Conceptually, one would expect that most negative social impacts – including issues of crime and family related problems – are a result of problem or excessive gambling in one form or another. In many respects, negative impacts from gambling consumption parallel those that result from alcohol consumption. For both commodities:

- A high proportion of the population participates as consumers at some time or another.
- A subgroup of all consumers are avid (heavy) consumers of the commodity.
- Within that subgroup there is some proportion of the total who consume to excess, to the point where their consumption patterns become a significant negative influence on their lives and the lives of those to whom they are responsible.

Problem/pathological gamblers thus parallel heavy drinkers/alcoholics, and create difficulties not only for themselves but also for other members of the community and society at large. A conceptual understanding of problem/pathological gambling is far from complete (Shaffer, Hall and Vanderbilt 1997). The empirical dimensions of problem gambling are even more difficult to achieve with any degree of reliability. Measuring prevalence of problem gambling is more difficult than measuring the prevalence of problem drinking, because there are fewer – if any – physical correlates. Problem gambling, to the extent it exists, is a largely invisible phenomenon. Measurement devices commonly used in prevalence studies, such as the South Oaks Gambling Screen (SOGS), were initially designed as clinical diagnostic tools rather than survey instruments, and should therefore be used with caution.

Going to the next stage – linking the prevalence of problem gambling to explicit social costs, such as costs associated with reduced productivity, absenteeism, unpaid gambling debts, criminal justice costs, social welfare costs, etc. – is even more difficult. To undergo such an exercise and come up with apparently precise aggregated cost estimates that can then be plugged into a cost–benefit type of framework, is either academically naïve or academically dishonest, or both. Thus, claims made by some authors that the costs to society from each compulsive gambler ranges from $15,000 to $35,000 per annum simply do not have either conceptual or empirical evidence to sustain them.

Measurement problems notwithstanding, there are legitimate concerns regarding the relative benefits and costs of permitted casino-style gambling as a function of the kinds of protections that regulatory constraints provide. Especially if consumer well-being (consumer surplus) dimensions are not given much priority, the policy trade-off is typically between economic benefits and social costs. Most of the economic benefits from casino-style gambling are attributable to tourism in one way or another.

Consider the following alternative frameworks for permitting casinos or casino-style gambling:

- destination resort casinos located away from population centres;
- casinos in rural (non-urban) locales within reach of population centres;
- casinos in urban or suburban locations in metropolitan areas;
- casino-style gaming permitted in a wide range of neighbourhood locations;
- gambling at home.

As we work down this list, the ratio of benefits to costs steadily declines. For destination resorts and for casinos in rural locales not too distant from population centres, casino gaming is exported to residents of other areas (i.e. tourists), and negative social impacts are exported as well. Such jurisdictions are net beneficiaries of economic injections that result in multiplier effects that are job creating and growth inducing.

With urban/suburban casinos, the ratio of benefits to costs declines relatively because the export component largely disappears; this is due to the fact that most of the casinos' customers are drawn from the metropolitan area. Social impacts stay within the area as well. Jobs created in the casino industry are, *ceteris paribus,* matched by reductions in jobs elsewhere in the metropolitan area, reflecting the shifting of spending patterns (Leven and Phares 1998). However, depending on expected profitability, such casinos can be important magnets for financial capital for construction and reinvestment.

Casino-style gaming permitted ubiquitously throughout a region – such as in bars and taverns – generates even lower scores on benefit/cost comparisons. This is because such authorization would trigger little in the way of new capital investment (especially in comparison to permitted casinos), and would likely have incrementally greater social impacts because of the increased access to potential consumers. It is possible that such casino-style gaming might be an efficient tax collector, however.

Gambling at home – via the Internet, interactive television, or phone betting – would show the lowest benefit/cost ratio. There would likely be little or no economic spin-offs into the local community. Social impacts could be significant in light of the high degree of access such wagering opportunities would create.

Though one might expect that there should be an increase in crime in areas that introduce casinos, the evidence to support such a hypothesis is neither very strong nor conclusive in its directions (Margolis 1997). Though, *ceteris paribus*, one might expect crime rates to increase as a result of problem gambling, this might be offset by improved job opportunities for previously unemployed workers. Furthermore, local tax receipts might lead to increases in funding for local law enforcement. That, coupled with increased security provided by casinos themselves, might push crime out of the casino's vicinity or the political jurisdiction where the casinos are located.

The issue of organized crime and systematic corruption with respect to casinos has steadily diminished over the past three decades. With few exceptions, modern permitted casino gaming in most countries has significantly limited any role of organized crime (there are some exceptions, such as Macao). This is mainly a result of effective and diligent regulatory regimes, along with new technologies that have made monitoring money within casinos easier to achieve.

One must keep in mind, however, that the commercial gaming industries are among the largest industries created primarily through the process of legislative permission and, as such, are subject to considerable lobbying efforts, or *rent seeking* (Walker and Barnett 1997). Legislative bodies may be more persuaded by arguments based on who will benefit from the economic proceeds that can come from legalization of casino-style gambling than by balancing the trade-offs between economic gains and social costs. Because the economic stakes can be so high, the potential for political corruption, or misplaced policy development, cannot be ignored.

THE FUTURE FOR CASINO GAMING

In terms of public perception, casinos have quickly been transformed from pernicious dens of iniquity to major catalysts for addressing a wide variety of economic concerns. With the exception of Asia, most regions of the world have softened their stance on prohibition against casino gambling, and have instead attempted to use casinos to address various economic objectives, including tourism development, economic development, tax revenue generation, job creation, foreign exchange enhancement, and combating illegal gambling operations.

The clearest regional or national objective when considering casinos has been tourism development, primarily because of the phenomenal growth of Las Vegas in the 1980s and 1990s. However, Las Vegas has set a somewhat unattainable ideal because of its critical mass of destination resort casino complexes; the fact that it had a legal monopoly on casino gaming in America for so long; its infrastructure, especially the airport and freeway linkages; and its geographic position in the western United States – close to Los Angeles and the large population base of Southern California, but not too close. Other jurisdictions may desire to emulate the tourism successes of Las Vegas, but it is unlikely that any will approach Las Vegas, either in size and diversity, or in proportion of customers who are indeed tourists.

Furthermore, if jurisdictions choose to legalize urban casinos as well as destination resort casinos, then much of the potential demand for casino-style gambling will be siphoned away from the more tourist-oriented resort casinos by the more convenient urban ones. In order for destination resort casinos to prosper, significant limitations need to be placed on casinos and casino-style gaming opportunities in population centres. Urban casinos and slot machines in bars and taverns (slot route operations) tend to draw a high proportion of their customers from the local area and thus, by definition, are not major tourist facilities.

On the other hand, there are clearly economic and demographic trends in the industrialized world that strongly suggest a bright future for casino-style gaming industries, as long as legitimacy and political acceptance are present (Simonson 1998). Discretionary spending on entertainment-based activities has grown significantly in the 1980s and 1990s, and casino gaming has proven to be a popular activity for those who enjoy thrill-seeking, but have grown too old to pursue physically challenging endeavours such as skiing or bicycling.

However, the proliferation of casinos has not occurred without detractors. In the United States, political controversy has followed nearly every state's attempt to introduce casino gaming in the 1990s. Various states, including Florida, Ohio, Pennsylvania and Texas, rejected substantial efforts to legalize casino gaming. A grass-roots coalition known as the National Coalition Against Casino Gambling led a strong backlash movement that was successful in persuading Congress to establish a National Gambling Impact Study Commission in 1997. Concerns about the morality of gambling, its effects on political corruption and criminal activity, and the spectre of pathological gambling and the damage it creates in its wake, have all been by-products of the public discussions that have surrounded the casino question.

There has been a clear dynamic at work with the spread of gambling. It has been legalized primarily as a means to achieve other 'higher purposes' such as tourism development, partly because the activity itself is considered – especially by policy makers – to be of questionable merit. Justifications are not found in the demand for gambling from the general public, but rather in the economic spin-offs that are thought to occur when gambling is authorized, especially in markets where it is prohibited.

When casino gambling only existed in destination resort areas – Las Vegas, Monte Carlo, the Caribbean, Atlantic City – associated problems were buffered by distance from where most customers resided. However, trends of the last ten years have resulted in many jurisdictions in urban and suburban casinos as well as slot machines widely placed in neighbourhood social gathering places. For better or worse, this has brought very attractive – and seductive – gambling very close to where many people live.

It remains to be seen whether such forms of casino-style gambling are a good idea overall, as the economic and social impacts are quite different among the various approaches. Nonetheless, economic pressures for expansions of the franchise to offer gambling will continue. Possible beneficiaries from newly legal gambling – Indian tribes, racetracks, bar owners, desperate cities, governments, declining tourist areas – will all push for the economic rewards that permitted gambling might bestow. However, what might work for any one of them cannot work for all of them, and many will be disappointed even if they get what they ask for, and especially if they *all* get what they ask for. In economists' jargon, economic rents can only exist as long as the supply of the commodity is constrained. If it is offered in a highly competitive environment, excess profits or rents are bid away.

This implies the ultimate winners in the spread of gambling might be consumers, but only if the negative externalities associated with gambling remain somewhat insignificant in the aggregate. If problem gambling really is as substantial as gambling opponents suggest, then the current course of expansion will lead to future regrets and perhaps future prohibitions. But if problem gambling is only a small part of the picture, or if it can be mitigated to the point that society will tolerate its dimensions, then the current directions are only problematic for the economic winners and losers, as is the case with any other commodity.

References

Campbell, Colin (ed.) (1994) *Gambling in Canada: The Bottom Line*, Vancouver: The Criminology Research Centre, Simon Fraser University.

Campbell, Colin, and Lowman, John (eds) (1990) *Gambling in Canada: Golden Goose or Trojan Horse?*, Vancouver: Simon Fraser University.

Christiansen, Eugene (1997) 'The United States 1996 Gross Annual Wager', supplement to *International Gaming and Wagering Business Magazine*.

Eadington, William R. (1995) 'The emergence of casino gaming as a major factor in tourism markets: policy issues and considerations', pp. 159–86 in Richard Butler and Douglas Pearce (eds) *Change in Tourism: People, Places, Processes*, London: Routledge Kegan Paul.

—— (1998a) 'Casino gaming–origins, trends and impact', pp. 3–15 in Klaus Meyer-Arendt and Rudi Hartmann (eds) *Casino Gambling in America: Origins, Trends and Impacts*, New York: Cognizant Communications.

—— (1998b) 'Contributions of casino style gambling to local economies', *Annals of the American Academy of Political and Social Sciences* 556 (March): 53–65.

—— (ed.) (1990) *Indian Gaming and the Law*, Reno: University of Nevada.

Eadington, William R. and Cornelius, Judy A. (eds) (1997) *Gambling: Public Policies and the Social Sciences*, Reno: Institute for the Study of Gambling and Commercial Gaming, University of Nevada.

Economic Development Review (1995) Special issue on gambling, Fall.

Gaming Board of Great Britain (1997) *Report of the Gaming Board for Great Britain, 1996/97*, London: The Stationery Office.

Goodman, Robert (1995) *The Luck Business*, New York: Free Press

Grinols, Earl, and Omorov, J.D. (1997) 'Development or dreamfield delusions: assessing casino gambling's costs and benefits', *Journal of Law and Commerce* 16 (1): 49–87.

Harrah's Entertainment, Inc. (1997) 'Harrah's survey of casino entertainment', Memphis Harrah's Entertainment, Inc.

Johnston, David (1992) *Temples of Chance: How America Inc. Bought Out Murder Inc. to Win Control of the Casino Business*, New York: Doubleday.

Journal of Travel Research (1996) Vol. XXXIV, Special issue on gambling, Number 3, Winter.

Kelly, Joseph (1993–94) 'Indian gaming law', *Drake Law Review* 43: 501–45.

—— (1997) 'The American Insurance Institute, like that bunny that keeps on going and going . . .', *Gaming Law Review* 1 (2): 209–12.

Kent-Lemon, Nigel (ed.) (1996) *World Casino Industry Review 1996*, London: Shiny International.

Kindt, John W. (1998), 'Follow the money: gambling, ethics, and subpoenas', *Annals of the American Academy of Political and Social Science*, 556: 85–97.

Leven, Charles and Phares, Don (1998) 'The economic impact of gaming in Missouri', a study presented to Civic Progress, St. Louis, Missouri.

Margolis, Jeremy (1997) 'Casinos and crime: an analysis of the evidence', Washington DC: American Gaming Association.

Meyer-Arendt, Klaus and Hartmann, Rudi (eds) (1998) *Casino Gambling in America: Origins, Trends and Impacts*, New York: Cognizant Communications.

Report of the Gaming Board for Great Britain, 1996/97 (1997) London: The Stationery Office.

Salomon Smith Barney (1998) *Quarterly Gaming Statistical Abstract*, Issue 1997, Fourth Quarter.

Shaffer, Howard J., Hall, Matthew N., and Vanderbilt, Joni (1997) 'Estimating the prevalence of disordered gambling behavior in the United States and Canada: A meta-analysis', Boston: Harvard Medical School.

Simonson, Robert J. (1998) 'Recreation, leisure, and gaming expenditures: exceptional long-term growth prospects', pp. 1–34 in William R. Eadington and Judy A. Cornelius (eds) *The Business of Gaming: Economic and Management Issues*, Reno: Institute for the Study of Gambling and Commercial Gaming, University of Nevada.

Thompson, William N., Gazel, Ricardo, and Rickman, Dan (1997) 'Social and legal costs of compulsive gambling', *Gaming Law Review* 1 (1): 81–9.

US General Accounting Office, May 1997 *A Profile of the Indian Gaming Industry*, GAO/GGD-97-91, Washington DC.

Walker, Douglas M. and Barnett, A.H. (1997) 'The social costs of legalized casino gambling reconsidered', paper presented to the 10th International Conference on Gambling and Risk Taking, Montreal.

9

TOURISM DEVELOPMENT AND NATIONAL TOURIST ORGANIZATIONS IN SMALL DEVELOPING COUNTRIES

The case of Samoa

Douglas G. Pearce

INTRODUCTION

Over the last two decades, the expansion of tourism in developing countries has generated a considerable amount of attention, debate and interest. Much research has focused on the diverse impacts which the growth of international tourism in developing countries has brought about (see Chapters 13 and 14) and on the way in which this development has occurred. Particular attention has been given to two usually divergent and sometimes conflicting sets of actors: multinational corporations and indigenous local participants (e.g. Smith and Eadington 1992; Pearce 1995). The emphasis has been on the private rather than the public sector. Research on government involvement has largely focused on tourism and public policy rather than on the agencies through which such policy has been formulated and implemented (Jenkins and Henry 1982; Pearce 1989; Holder 1992). This reflects a wider pattern of neglect as it is only in the 1990s that tourist organizations *per se* have been subjected to systematic scrutiny and analysis by researchers (Pearce 1992; Choy 1993). Choy considered the top five Asia–Pacific destinations, but the majority of the research in this field so far has been concerned with the role and structure of tourist organizations in developed Western economies, leaving those in developing countries essentially unexplored. A recent review of tourism in the South Pacific showed, for example, that while national and other tourist organizations in Australia and New Zealand were attracting increased attention, with the exception of Sofield's (1991) examination of the organizational aspects of the *naghol* in Vanuatu, little comparable institutional research appears to have been carried out yet in the Pacific islands (Pearce 1995).

Comparative research and single-country studies show tourist organizations come in many shapes and forms; they may have a variety of economic, social and other goals and they may be monofunctional or multifunctional, that is they exercise one or more functions such as marketing, development, planning, research, visitor servicing, lobbying or regulation (Pearce 1992, 1996a and b, 1997a; Choy 1993). Their general purpose is to foster the growth of tourism through the leadership they exert and co-ordination they provide of a multifaceted sector of the economy. While common patterns and tendencies have been identified, this author has concluded (Pearce 1992: 200) that: 'There is no single best type of [tourist] organization nor interorganizational network, rather each country must evolve a system which best reflects local, regional and national conditions.'

This chapter seeks to extend this field of research through a detailed case study of one South Pacific national tourist organization, the Samoa Visitors Bureau (SVB). The aim is to examine the structure and functions of a national tourist organization (NTO) in a small developing country and to assess its role in the light of the tourism development policies being pursued there and in terms of broader contextual factors. In these respects it follows an approach developed for the study of tourist organizations elsewhere in which inter-organizational analysis is used to identify and evaluate differing goals and functions in a tourist organization network and to assess these with respect to the broader environment or general contextual factors, a sub-environment of which is the tourism environment (Pearce 1992). Although relationships with other partners form a key part of the research, the analysis is not inter-organizational in the sense that the previous studies were, as no network of official tourist organizations exists in Samoa due to its small size and the relative youthfulness of its tourist industry. However, the interaction between the NTO and these and other environmental factors forms a central dimension of the open systems analysis and may serve to illustrate both how such research may be undertaken in other small developing countries and what issues may arise there.

The chapter is based on a wide-ranging review of the SVB undertaken by the author in December 1996 as part of the New Zealand Official Development Assistance (NZODA) tourism programme in Samoa (Pearce 1997b). The review was undertaken from two perspectives: internal and external. Internally, consideration was given to evaluating how the SVB was structured and functions and how its role and activities were perceived by the general manager and by each of the SVB's divisional managers and overseas representatives. Frank and in-depth interviews with management and overseas representatives were complemented by analysis of relevant documentation (e.g. annual plans and reports, statutory material, promotional literature) and by participant observation. In this latter regard the author was invited to participate in a meeting of the National Beautification Committee, attend a tourism seminar, accompany a SVB team on its meetings with two village councils (at Salamumu and Lefaga) and make site visits on both Upolu and Savaii. The external appraisal of the SVB and its activities was obtained through discussions with selected tourist industry representatives in one of the major markets (New Zealand) and on both Upolu and Savaii. The candour and openness of these discussions made possible a realistic assessment of the linkages between the SVB, and industry partners and provided insights into how the organization is perceived from outside. Finally, information from these diverse sources was synthesized and evaluated.

CONTEXTUAL FACTORS

Samoa is a small multiple island state in the South Pacific which gained independence from New Zealand in 1962 (Figure 9.1). Formerly known as Western Samoa, the country took the name Samoa in 1997. This term is used throughout this chapter and applied to the country's NTO, even if at the time the research was undertaken it was still the Western Samoa Visitors Bureau. Samoa consists of two main islands – Upolu and Savaii – and five smaller ones, covers an area of 2,935 km^2 and has a population of 163,000, of whom almost three quarters live on Upolu and a fifth in the capital, Apia. Some 80 per cent of all land is held in customary tenure; communal rights to and use of land by extended family members underpins Samoan society and culture (Government of Western Samoa/TCSP 1992). The country's economy has been based on semi-subsistence agriculture, with some plantation production and limited forestry and light manufacturing. Fairbairn-Dunlop (1994: 127) states:

> Like other similar Pacific nation states, Samoa's economic development is constrained by factors such as its small population and workforce, limited land area and resources, isolation from major trade routes, restricted range of marketable products and limited number of markets and suppliers. In this context tourism, although small in relation to other destinations, is now making a significant contribution to the economy of Samoa. Estimates by the SVB and Central Bank of Samoa for 1995 indicate that tourism generated 86.2 million tala (cf. 46.6 million in 1990), accounted for 25 per cent of GDP and 32 per cent of foreign exchange earnings (WSVB 1996).

Tourism in Samoa is still at a very youthful stage of development and despite significant recent increases tourism remains small scale, both in absolute and relative terms. Total visitor accommodation in 1996 (not including open beach fales) amounted to 740 rooms in thirty-six establishments. The

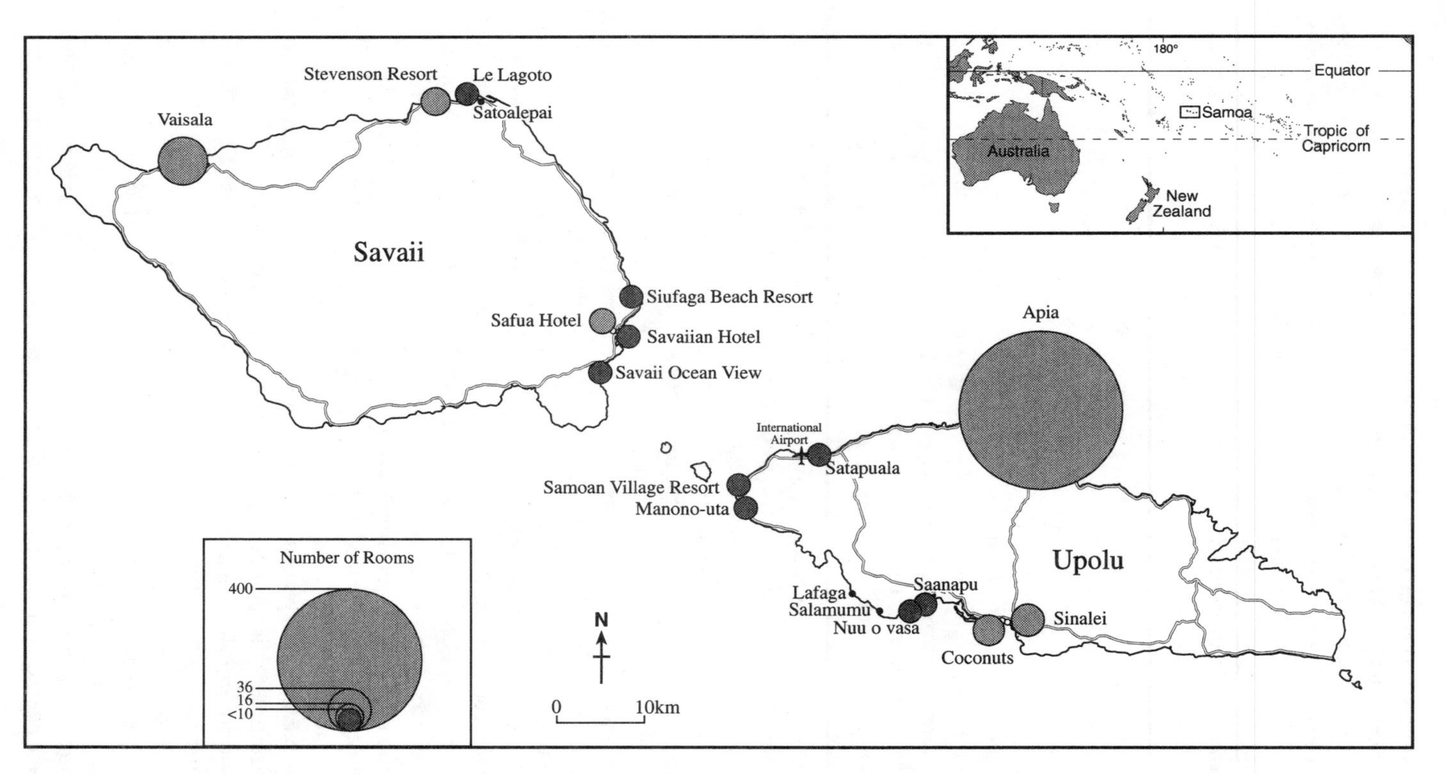

Figure 9.1 Samoa: distribution of accommodation (1995) and place names.

Source: Samoa Visitors Bureau.

large majority (86.8 per cent) of the rooms are located on Upolu (of which 70.5 per cent are in Apia), with only 13.2 per cent being found in Savaii. Only four of the properties have more than fifty rooms; twenty-two have less than fifteen. Most have been built since 1990. Few proprietors and managers entering the accommodation sector have had previous experience or background in the hospitality industry. Overall, the accommodation sector might thus be characterized as small, recent, generally unsophisticated and lacking professionalism. Other sectors, with the exception of the major carriers, tend to share these characteristics, although important advances are being made in the field of inbound tour operations.

Significant increases in visitor arrivals to Samoa have been experienced in recent years: 1993, 47,071; 1994, 50,144; 1995, 68,392; 1996, 73,155. It should be noted, however, that visitor arrivals had previously peaked at 53,994 in 1989, the downturn in the intervening years being largely attributable to the effects of the cyclones in 1990 and 1991. The four leading sources of arrivals in 1996 were: American Samoa (33 per cent), New Zealand (28 per cent), the USA (11 per cent) and Australia (10.6 per cent). Holidaymakers accounted for only a 30 per cent of total arrivals, behind those visiting friends and relatives (36 per cent) and ahead of business travellers (12 per cent) and other visitors (22 per cent). In absolute terms, a total of 22,289 holiday arrivals were recorded in 1996. Thus, while the number of visitors to Samoa has been increasing, the scale of visitation at present is small. The country ranks as a middle-order South Pacific destination well behind the regional leaders, Fiji and French Polynesia (Pearce 1995).

The economic significance of tourism has increasingly been recognized in recent years by the Government of Samoa. While the Government recognizes the economic contribution that tourism makes it is also concerned that a balanced approach to the growth of tourism is adopted. To achieve this, a ten year plan, the Tourism Development Plan 1992–2001 (Government of Western Samoa/TCSP 1992), was prepared in 1992 and has subsequently guided the growth of the sector in Samoa. The country's official tourism development policy as outlined in the country's Sixth Development Plan 1988–90 can be summarized as follows:

> The basic objective of tourism development will be to ensure that development is achieved without causing undue strain on the economy, while ensuring that the well-being of society is not jeopardised through the social changes that tourism development sometimes entails.
>
> (Cited by Government of Western Samoa/TCSP 1992: 50–1)

Specific tourism development objectives outlined in accordance with this policy and which remain current (WSVB 1996) are:

- to generate employment and increase local income though providing greater opportunities for active local participation in tourism development;
- to enhance and preserve Samoa's rich cultural heritage and beautiful natural environment, both of which provide important types of tourist attractions;
- to promote the expansion of other economic sectors through developing and improving the linkages with those sectors;
- to retain a higher share of tourist expenditures in the local economy;
- to increase the level of tourism awareness amongst local people; and
- to promote and increase visitor interest in the traditional and potential source markets.

Constraints on the future development of tourism in Samoa identified in the development plan, subsequent documents and discussions during the course of the review, include:

- customary land tenure issues
- questions of air access
- insufficient accommodation outside of Apia
- lack of investment capital within Samoa
- competition from other South Pacific destinations

These constraints are inter-related, generally well-recognized and unlikely to be solved in the short term.

The *faaSamoa*, the traditional Samoan way of life, is also a major factor influencing the nature and extent of tourism development in Samoa. Samoan villages are well-structured entities made up of a number of *aiga* (extended family units), each of which has an appointed chief or *matai*. The *matai* come together in a council of chiefs which serves as the village's law-making and decision-making body. The village women's committee also plays a major role in village affairs. Fairbairn-Dunlop (1994: 124) describes the *faaSamoa* as:

> . . . a system of chiefly rule in which every person is expected to know their place and the correct behaviour patterns of their place. 'Correct behaviour' is the dynamic which ensures the smooth running of the chiefly system, the norm being that one gives service and respect to those in higher status and can expect to receive the same from those of lower status.

She continues (p. 125): 'In the *faaSamoa*, prestige is achieved by giving – by sharing rather than accumulating resources. Hence, the system is marked by a continuous exchange of goods and services.'

Fairbairn-Dunlop notes that the *faaSamoa* provides Samoan women with the opportunity to become actively involved in tourism development and provides examples of successful individual entrepreneurs and the role of the women's committees. More generally, the 1992 tourism development plan recognizes the significance of diverse aspects of the *faaSamoa*, such as the forum which the council of chiefs may provide for discussions on village-based tourism and the opportunities for use and abuse which Samoan 'expressive hospitality' may offer both villagers and visitors.

ESTABLISHMENT, EVOLUTION AND FUNCTIONS OF THE SAMOA VISITORS BUREAU

The SVB is a statutory body set up in 1986, following passage of the Western Samoa Visitors' Bureau Act 1984. The new autonomous organization succeeded the small (1980 staff of four) Visitors Bureau Section under the jurisdiction of the Department of Economic Development (Margraff 1980). As laid out in the Act, the functions of the SVB are:

(a) to encourage visits to Samoa by people from other countries, and travel and holidays within Samoa by people from Samoa or other countries;
(b) to encourage the development and improvement of facilities for tourists;
(c) to co-ordinate the activities of persons providing services for tourists;
(d) to advise the Government on all matters relating to tourism; and
(e) to prepare, implement and maintain a comprehensive tourism development plan for Samoa.

As might be expected in the evolution of any new organization, the SVB has grown and undergone structural changes in its first decade. In April 1992, at the time of the preparation of the development plan, the SVB had a staff of thirteen, including five clerical and general positions. By December 1996 the staff had more than doubled to twenty-eight, while its annual grant from Government had more than trebled to three million tala for 1996/97. As is illustrated by Figure 9.2, the SVB is currently structured into five divisions, each directly responsible to the general manager: marketing and promotions, planning and development, education and training, research and statistics, and finance and administration. This structure reflects certain recommendations in the 1992 plan and other changes implemented by the current general manager. Significant recent developments include the establishment of overseas offices in Sydney (May 1993) and Auckland (February 1996). In summary, as a national tourist organization, the SVB today might be characterized as small (especially in terms of its budget), youthful and multi-functional.

In terms of resource allocation, marketing and promotions is clearly the most well-endowed division, receiving almost 50 per cent of the total

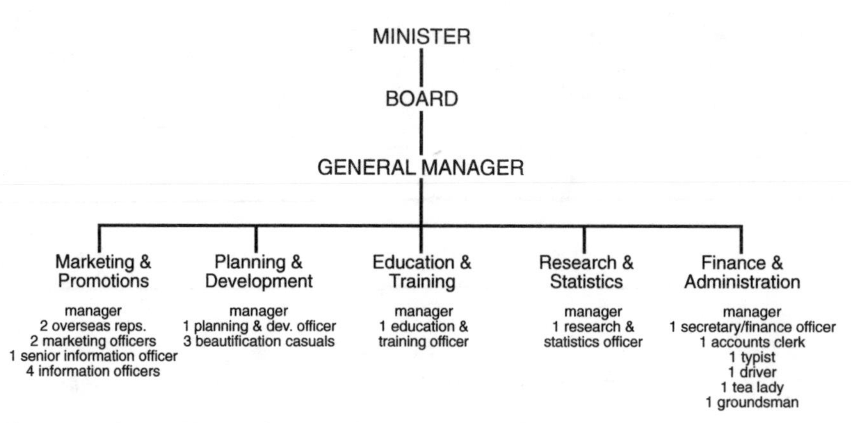

Figure 9.2 Samoa Visitors Bureau: Organizational Chart.

budget and almost 80 per cent of development expenditure in the current year. Nevertheless, the sums involved are small relative to the markets being tapped and the competition in these. While the SVB's marketing division undertakes its own activities such as the organization of familiarization trips and wholesaler and retailer contact and support in the New Zealand and Australian markets, it also stretches its resources through joint participation in trade and consumer shows with other South Pacific NTOs. For instance, the SVB participates in the major European trade fairs in association with and with the support of the Tourism Council of the South Pacific, and its Auckland representative is an active member of SPANTO (South Pacific Association of National Tourist Organizations).

This marketing thrust is a common characteristic of NTOs and in many developed countries international marketing has become the dominant or even sole function of such bodies (Pearce 1992; Choy 1993). The SVB is distinguished by the diverse range of functions currently undertaken, especially the relative emphasis given to planning and development and education and training, each of which is a separate division. This emphasis reflects the incipient stage of development of the sector in Samoa and Government policies towards the balanced and orderly development of tourism.

In recent years, and with the support of donor aid programmes which have encouraged small scale projects, notably those of the NZODA and AusAID, the SVB, through the managers of the education and training and planning and development divisions, has been implementing a strategy to foster local participation in the development process and maintain the *faaSamoa*. This strategy has several strands. Firstly, villagers' awareness of the nature and potential of tourism has been heightened through a series of village tourism workshops in which ways of becoming involved in tourism are outlined to the council of chiefs and advisory services are offered. This may be followed by the provision of grants (e.g. for water supply or toilets), hotel staff or guide training and marketing support. In the case of some of the small resort developments SVB staff have 'softened the ground' for the project, facilitating discussions on land issues, the need to ensure the security of guests, giving priority to the employment of staff from the local village and their training. A more general awareness of tourism has also been created though a regular talk-back radio programme and television appearances. Infrastructural assistance is being provided by way of a village signage programme, moves to establish scenic viewpoints and provide additional attractions through, for example, archaeological restoration programmes. In addition, the SVB co-ordinates the activities of the National Beautification Committee which was set up after the devastating hurricanes of the early 1990s and which is playing a leading role in planting and landscaping in Apia and fostering village pride and appearance through the village beautification programme. Largely by default, the Beautification Committee, and thus the SVB, also becomes involved in other activities less directly related to the development of tourism.

It is too early yet to evaluate fully the impact and fruits of these activities. Early grants providing infrastructural assistance have not always led to sustained development in the short to medium term. Insufficient attention has occasionally been paid to ongoing maintenance leading to the run-down of facilities and in parts of Upolu there would appear to be an excess development of beach fales. The broader approach now being pursued of initiating development through the village workshops would seem to offer much more potential than the provision of grants alone. The wetlands ecotourism project at Satoalepai supported by financial assistance from AusAID and input from the SVB has the potential to become a successful community-based tourism venture providing a range of complementary activities offering additional attractions to visitors and sources of income to the villagers can be incorporated with the outrigger-canoe trip and an appropriate management system can be put in place. The village beautification programme would also appear to be contributing to the attractiveness of villages on Upolu and Savaii as well as to enhancing local community pride. Certainly to date the SVB's success in all these areas should not be overstated but the actions of the SVB would nevertheless appear to be a useful means of attempting to go beyond the mere rhetoric of advocating local participation in tourism.

In other areas the SVB's role in the maintenance of local culture and the development of tourism might be questioned. In recent years the SVB has become increasingly involved in the organization and day-to-day running of major events such as the Musika Extravaganza, the Teuila Tourism Festival and the Miss Samoa Pageant. These events place considerable demands on the SVB and detract from its core activities. While the events may benefit local communities, discussions with tourist industry representatives suggest that so far only minimal new demand is being created. As with some aspects of the National Beautification Committee, the SVB appears to have assumed responsibility for event organization largely by default and should now pass on the actual running of these to other relevant groups and agencies and limit its own role to some co-ordination and to enhancing the tourism spin-offs through appropriate marketing.

One key area where the SVB does have a key leadership and co-ordination role to play is in the development of partnerships with the individuals, businesses and agencies from both the public and private sectors who provide the many and diverse services and facilities used by tourists. NTOs frequently have a co-ordination function and undertake common good services such as destination marketing and research, but leadership and partnership building are particularly important in developing destinations such as Samoa given the characteristics of its emerging tourist industry outlined in the review of environmental factors. Here the SVB might be depicted as a pivotal organization with numerous partners (Figure 9.3).

The links between the SVB and these partners are many and varied. The organization of a familiarization trip by a visiting travel writer, for instance, may begin by an overseas marketing representative selecting an appropriate journalist, an on-island itinerary is arranged by SVB staff in Apia, with assistance in kind provided by a carrier and local hoteliers, restaurateurs and other operators. Or, as noted earlier, training and development assistance may be given by SVB staff to a village ecotourism project, with financial grants being made available from a donor agency through an aid programme managed by the NTO. Collective contact with the private sector members is provided by meetings with the Samoa Visitors Association and the Savaii Visitors Association. Internationally, the SVB has combined forces with and drawn on the resources of such organizations as the Tourism Council of the South Pacific (TCSP), the Pacific Asia Travel Association (PATA) and SPANTO.

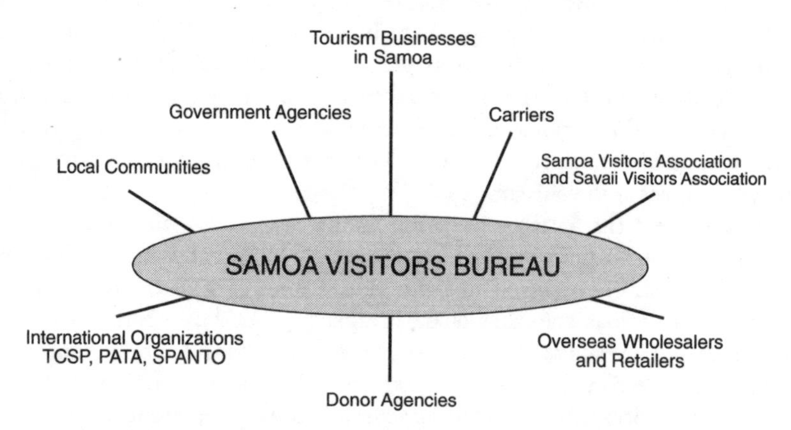

Figure 9.3 Partnerships of the Samoa Visitors Bureau.

The issues raised at the tourism seminar convened by the SVB in December 1996 and attended by sixty participants is indicative of the need for leadership and co-ordination and a reflection of the stage of development which tourism in Samoa is now at. One group focused on matters affecting the accommodation sector and addressed such issues as:

- the need to design and implement an appropriate classification system;
- the use of local currency, the Samoan tala, rather than American dollars;
- improved communications between the SVB and the industry, especially with regard to accommodation statistics;
- accommodation, plant and infrastructure e.g. problems relating to electricity, water supply and garbage disposal;
- the environment and sustainable tourism.

The second group considered issues relating to entry formalities and visitor arrival statistics while the third discussed a range of transportation and touring matters such as the abolition of the practice of stamping international drivers' licences, rental car regulations and entry fees to sites on customary land.

While much work is already being done in these and other areas the various linkages do not currently appear to form part of a big picture but rather consist of a series of separate activities. Scope exists for the SVB to develop its leadership role by building on the present rather loose and *ad hoc* network to strengthen these partnerships and make the links more visible and effective. Firstly, the NTO needs to recognize explicitly its partners and the need for synergy, and consciously strive to foster a collective approach to the development of tourism in Samoa. Secondly, the SVB needs to review in more detail the nature of its relationships with each of the partners shown in Figure 9.3 and examine ways of making these more effective.

Discussions held during the course of the review indicate that the SVB is generally viewed positively by its partners. However, criticism was commonly voiced by industry representatives with regard to a lack of feedback and follow-up. A common complaint, for example, was that those hosting visiting media often saw no specific outcome (e.g. magazine articles or published photos) from their efforts, with some being reluctant to offer their services in the future. Others bemoaned the lack of timely visitor statistics and the late notification of trade shows and other activities. Some of the Auckland wholesalers thought the SVB could be more proactive and provide greater brochure support. Hoteliers on Savaii perceived their island was getting much less attention than Upolu (this is not borne out by current promotional material). The visits to meet with the village councils in Salamumu and Lefaga also revealed a breakdown in communications.

Many of these shortcomings, real or perceived, might be overcome by improved communications between the SVB and its many partners. In some cases regular meetings will be appropriate, especially with the visitors' associations and the carriers. In others, information might be disseminated through, for example, monthly release of visitor statistics and quarterly newsletters heralding forthcoming events, reporting on developments in Samoan tourism and outlining the SVB's activities, particularly the outcomes of joint programmes in training and marketing.

DISCUSSION AND CONCLUSIONS

The current multifunctional nature of the Samoa Visitors Bureau has its parallels in other NTOs when tourism has been at a relatively early phase of its development and when marketing has been complemented by an active period of development assistance through such means as grants, incentives and the provision of advisory services (Pearce 1992). Before a destination can promote itself, it must first have developed a product to market. Choy (1993: 364) argued: 'It is unlikely that government can compensate for the lack of a strong private sector; and initial government efforts should be directed at fostering the expansion of the private sector in providing tourist services'. What is distinctive about the Samoan case is the way in which the country's strong social and cultural traditions – the *faaSamoa* – have been incorporated in government tourism policies and the ways in which these policies are being implemented through the NTO in their attempts to foster local participation in the development process. It is noticeable too that these attempts are being supported by overseas assistance programmes which have focused on small scale projects, the NTO acting as a sector specific intermediary through which external aid is being channelled. This is not to say that the construction of larger coastal resort complexes in Samoa has not been considered, debated or ruled out of the question (Government of Western Samoa/TCSP 1992; Pacific International Consulting Network 1995). Rather, the issues of customary land tenure have so far acted as a major constraint on this form of development.

Given current Government policies and that further growth in tourism would require even more effort to be directed at cultural and social matters, the 1996 review recommended that a new division of Samoan and Cultural Affairs be established within the SVB with specific responsibilities in this field, notably:

- liaison with the villages in regard to all tourism development matters such as land tenure, heightening tourism awareness and enhancing local participation;

- co-ordination of cultural facets of event organization and of performing and expressive arts where these have a significant tourism component;
- monitoring the cultural and social impacts of tourism so as to enhance Samoa's cultural heritage and mitigate any adverse impacts.

Such a division was established the following year.

The diversity of functions which the SVB has assumed or acquired in some instances also reflects the existence of what might be considered an institutional vacuum. The running of cultural events and some aspects of the work of the National Beautification Committee fall outside what might be considered the core business of an NTO but the SVB has picked up these functions in the absence of other bodies – Apia does not have, for instance, a municipal local authority. While such activities of these bodies may be important in themselves they may only be indirectly related to the growth of tourism, and the SVB's role in these areas needs reassessing.

In these ways the SVB currently is engaged in a very wide spectrum of activities, from active participation in community-based tourism through to direct representation in international markets, a very challenging situation in terms of both personnel and funding. In general, what is required now is the consolidation of the recent phase of the NTO's growth by sharpening the focus of the roles and functions of the SVB, making these more explicit and taking a longer-term perspective on the organization's development. In this regard, the organizational review recommended a number of key actions be taken. These included:

- development of a clear mission statement;
- redesignation as a tourism authority giving greater visibility to its diverse functions;
- development of the NTO's leadership role through strengthening partnerships in the public and private sectors.

In conclusion, while the SVB exhibits features of other national tourist organizations it also has a marked individual identity influenced by strong national social and cultural characteristics and the youthful stage of development of both the organization and tourism in Samoa. Such a finding is not surprising in view of the open systems approach adopted which has placed emphasis on the impact of contextual or environmental factors. Further comparative work is now needed in order to ascertain the degree to which the features described are specific to Samoa or common to other small developing countries. How, for example, are cultural considerations expressed in the actions of other NTOs? Are there other ways in which NTOs can foster local participation in tourist development? To what extent do the forms of external aid determine the way in which NTOs in small developing countries are structured and function or in what ways is overseas assistance conditioned by the NTO's policies and activities? Are institutional vacuums common? How are they filled elsewhere? By exploring these and other questions in a range of developing countries researchers will more readily be able to evaluate the role of national tourist organizations in the development process.

Acknowledgments

This paper draws on an organizational review of the Samoa Visitors Bureau undertaken by the author for the Samoa Visitors Bureau, Tourism Resource Consultants and the Ministry of Foreign Affairs and Trade as part of the NZODA programme for Samoa. Thanks are extended to all those who assisted with the review, especially Sonja Hunter, General Manager of the Samoa Visitors Bureau, and Dave Bamford of Tourism Resource Consultants. The author alone remains responsible for the views expressed in this paper.

References

Choy, D. (1993) 'Alternative roles of national tourism organizations', *Tourism Management* 14 (5): 357–65.

Fairbairn-Dunlop, P. (1994) 'Gender, culture and tourism development in Western Samoa', pp. 121–41 in V. Kinnaird and D. Hall (eds) *Tourism: A Gender Analysis*, Chichester: Wiley.

Government of Western Samoa/Tourism Council of the South Pacific (1992) *Western Samoa Tourism Development Plan 1992–2001*, Apia: Government of Western Samoa/Tourism Council of the South Pacific.

Holder, J.S. (1992) 'The need for public-sector cooperation in tourism', *Tourism Management* 13 (2): 157–62.

Jenkins, C.L. and Henry B.M. (1982) 'Government involvement in tourism in developing countries', *Annals of Tourism Research* 9 (4): 499–521.

Margraff, V. (1980) 'Research requirements of tourism in Western Samoa', pp. 163–8 in D.G. Pearce (ed.) *Tourism in the South Pacific: The Contribution of Research to Development and Planning*, New Zealand Man and the Biosphere Report No. 6, New Zealand National Commission for UNESCO/Department of Geography, University of Canterbury, Christchurch.

Pacific International Consulting Network (1995) *Tourism Investment Study*, Apia: Western Samoa Visitors Bureau.

Pearce, D.G. (1989) *Tourist Development*, 2nd edn, Harlow: Longman.

—— (1992) *Tourist Organizations*, Harlow: Longman.

—— (1995) Tourism in the South Pacific: patterns, trends and recent research, *Progress in Tourism and Hospitality Research* 1 (1): 3–16.

—— (1996a) 'Federalism and the organization of tourism in Belgium', *European Urban and Regional Studies* 3 (1): 189–204.

—— (1996b) 'Tourist organizations in Sweden', *Tourism Management* 17 (7): 413–24.

—— (1997a) 'Tourism and the Autonomous Communities in Spain', *Annals of Tourism Research* 21 (1): 156–77.

—— (1997b) *Western Samoa Visitors Bureau Organizational Review*, unpublished report prepared for the Western Samoa Visitors Bureau, Tourism Resource Consultants and the Ministry of Foreign Affairs and Trade.

Smith, V. and Eadington, W. (eds) (1992) *Tourism Alternatives: Potentials and Problems in the Development of Tourism*, Pittsburgh: University of Pennsylvania Press.

Sofield, T.H.B. (1991) 'Sustainable ethnic tourism in the South Pacific: some principles', *Journal of Tourism Studies* 2 (1): 56–72.

WSVB (1996) *Annual Work Plan 1 July 1996–30 June 1997*, Apia: Western Samoa Visitors Bureau.

10

SMALL SCALE ENTERPRISES IN THE TOURISM INDUSTRY IN GHANA'S CENTRAL REGION

William C. Gartner

INTRODUCTION

Much has been written regarding development on the African continent. A number of theories have been proposed to explain why western-based development models frequently fail. The two main theories, with most development experts assigning the other lesser accepted theories to one of these, are the neo-classical model, supported by the developmental or modernization school, and dependency theory, supported by the dependency school (Anunobi 1994).

The neo-classical school argues that development is stymied by poor or inadequate infrastructure, an uneducated and unskilled workforce, weak management skills, low or non-existent levels of savings and capital formation, and strained international relations. The dependency school argues that developing countries are in fact economic colonies of developed nations and suffer from poor trade terms both importing and exporting. The dependency school argues further that small enterprises are most severely affected by import liberalization programmes effectively crowding them out of the market. This is especially important in Africa where most of the businesses are classified as Small Scale Enterprises (SSEs) and most GNP is generated by SSEs.

Small Scale Enterprises were not extensively studied in Africa prior to their 'discovery' by the International Labour Office in the early 1970s (Ongile and McCormick 1996). Those examining the record before, and for a period of years following this 'discovery', concluded that very little is known regarding the composition and characteristics of this sector (Liedholm and Mead 1987). The first studies treated SSEs as qualitatively similar to Large Scale Enterprises. It is only of late that more extensive studies have been conducted leading to a greater understanding of how important SSEs are to Africa's economic future (McCormick 1992).

As can be imagined, the studies that have been completed focus on traditional occupations such as agriculture, textiles and trading. No studies focusing on tourism-dependent firms (e.g. hotels, gift shops, tour operators) were uncovered. This is not surprising as tourism is a rather new economic activity for many countries in Africa, plus donor agency money for the most part has not been directed at tourism development but instead has supported efforts focused on traditional industry.

SSEs are officially defined as having less than fifty employees (Hansom 1992). Certain qualitative characteristics have also been proposed to define the sector. Some of the more common include just-in-time production for direct sale of product to consumer, lack of specialization in the labour force, poor or non-existent bookkeeping, and heavy employment of family workers (Scheider and Barthold n.d. from Hansom 1992). It is estimated that the number of SSEs in many developing countries is far larger than the number of medium or larger firms and accounts for between 40 and 90 per cent of non-government employment.

It is the intent of this chapter to examine the development role of tourism-dependent SSEs, specifically the hotel sector, in Ghana's Central Region. In doing so, the discussion will focus on the role of financing these enterprises, including an examination of key indicator ratios, and finally an examination of one's social obligations, as determined by cultural affiliation, affecting hotel management.

A related issue involves the sustainability of the enterprises now in place. Much of the tourism development literature is in favour of local development utilizing local resources including capital. This is certainly the case in Ghana's Central Region where no multinational hotel corporation has yet begun to operate. One of the reasons for the absence of larger hotel properties is the newness of the destination. Historically large-scale enterprises do not operate in areas where profit is questionable and market instability exists (Pedersen 1997). Clearly, given the history of tourism development in Ghana, this is the situation facing all investors in tourism-dependent businesses in the Central Region. The question that must be considered is: If tourists to the area increase due to product development and a stable political environment, what is the future of the locally owned and operated small-scale operations now in place?

CONTEXT

The Republic of Ghana is located on the West Coast of Africa approximately 750 miles north of the Equator. A tropical country with temperatures ranging from 21 to 32°C, it is relatively moist in the south along the coastal zone with annual rainfall averaging 2,030 mm. Tropical forests at one time encompassed all of the southern and middle sections of the country. Few extensive tracts remain today. Ghana has substantial gold and, at one time, timber reserves and is also a major exporter of cocoa. In spite of its resource wealth, development has been hampered by a burgeoning population (estimated at 15 million) and a series of coups from independence through to 1981. It has been the coups, both attempted and successful, that have, more than anything else, retarded tourism development (Teye 1988).

Political stability since 1981 and the adoption of International Monetary Fund (IMF) guidelines have had a positive impact on many sectors of the economy including tourism. Tourism development is not new to Ghana, as during the mid 1970s there was substantial development of tourist products facilitated by the construction of some very large hotels throughout the country. What is new about the recent level of tourism development is the focus on private-sector ownership of the industry. Where in the 1970s almost all tourism development was conducted by government, supported by high earnings for cocoa that all came crashing down as the world price of cocoa collapsed, recent growth is being fuelled by private initiatives. The tourism growth trend is evident by construction of new four and five star hotel developments throughout the capital city of Accra. The rural situation is equally promising as a major tourism development project in Ghana's Central Region is leading the wave of rural tourism development.

Ghana is now governed by an elected president and Parliament. Almost all Members of Parliament are from the president's party, which often leads to charges that democracy does not work for Ghana. The country has, nevertheless, experienced economic growth, especially since the start of the 1990s, and tourist numbers do show an increase. Recent figures indicate total arrivals now exceed 250,000 per annum, up from 100,000 in 1990. Arrivals are dominated by business travellers. Although accurate figures are hard to come by, it is estimated that only 30 per cent of all arrivals could be classified as pleasure tourists with the remainder arriving for some business transactions or study trips. Donor agency activity is high in the 1990s compared to what existed in the 1980s and it is argued that much of the increase in arrivals is due primarily to spending by the various donor agencies operating in Ghana. This argument is further supported by figures which reveal that almost two-thirds of the visitors (64.3 per cent) to the country are travelling alone, high repeat visitation (54.5 per cent), and very low utilization of local tour operators' services (2.5 per cent) (Ministry of Tourism 1996).

Ghana, with assistance from the United Nations Development Programme and the World Tourism Organization, has recently completed a tourism master plan (Ministry of Tourism, Ghana 1996a). All regions in the country were assessed in terms of their tourist-generating potential. As expected the Greater Accra region, where the capital and major urban centre are found, is expected to receive the most arrivals but the Central Region, assisted by the development project briefly outlined in this chapter, is one of the few regions where tourist attracting resources in a developed state can be found. Because of this it is a major part of the tourism master plan, which estimates tourist arrivals to the country will increase to over one million by 2010.

Ghana's Central Region, approximately two hours' drive by car from Accra, with its historic capital of Cape Coast as the central point is the site of the first European contact with West Africa (Figure 10.1). The epicentre of this contact is Elmina Castle (Fort George Castle under British rule) which was originally a small fort erected by the Portuguese in 1482 to facilitate contact and trade with West African indigenous peoples. The fort later grew into a major trading centre and fortification to protect European interests on the continent. Elmina Castle is one of twenty-six European fortifications still standing along Ghana's Coast. Both Elmina and the nearby Cape Coast castle are on the World Heritage list of significant historical buildings. One of the most valuable, in economic and sociocultural terms, commodities to pass through these castles was people–slaves. It is estimated that millions of slaves were processed through these castles and today the journey of the slaves from their homes to the castles is the basis for the World Tourism Organization's Slave Route project.

Significant other resources exist in the Central Region including small tropical forest reserves which hold great biodiversity importance since they once were part of the great, isolated West Africa tropical rain forest belt. The region's cultural resources are another major tourist attraction as parts of the culture are readily shared with visitors.

Since the collapse of cocoa prices in the 1970s, and impacts from subsequent *coups d'état* the Central Region's industry in general and the tourism sector in particular were almost non-existent. The only reason some hotels were still operating was to serve the number of out-of-region visitors attending weekend funerals (see below). In 1988 a group of government representatives associated with the Central Region Development Commission (CEDECOM) developed a plan to restore Elmina and Cape Coast Castles and develop a forest reserve as a national park. This plan was then presented to donor agencies resulting in a $5.6 million development project for the region funded by the United States Agency for International Development. The development consortium that was assembled consisted of Conservation International, responsible for national park development; the Smithsonian Institution, responsible for museum development and interpretive services; an architectural firm from the United Kingdom, responsible for castle restoration; and the University of Minnesota Tourism Center, responsible for promotion, marketing and private and public sector training. Today the project has been successful in restoring and stabilizing both castles, creating Kakum National Park which has the only canopy walkway on the African

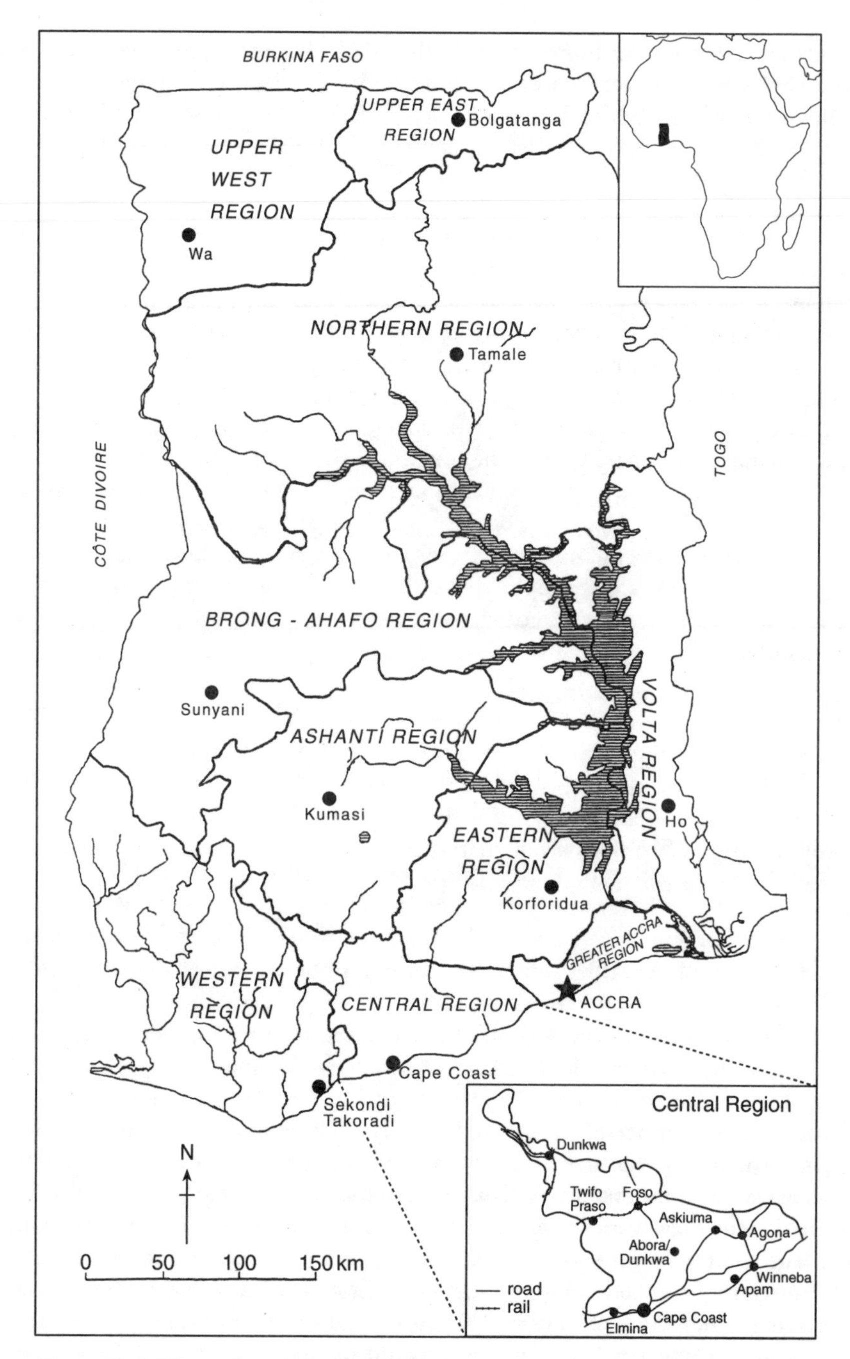

Figure 10.1 Ghana: location and place names.

Source: After National Tourism Development Plan for Ghana.

continent, developing museums in both castles and an interpretive centre at the park's headquarters, and organizing and training large segments of the growing tourism industry. It is the organizing and training of the private-sector tourism interests that led to the research for this chapter.

RESEARCH METHODOLOGY

The research conducted for this chapter was part of a larger study of a microeconomic assessment of the tourism industry in the Central Region centred around the Cape Coast/Elmina area. The Cape Coast environ was chosen as the focus area because all the development to date for the project described above is in close proximity to the sister cities of Cape Coast and Elmina. They form the largest city complex in the region.

The microeconomic assessment had both qualitative and quantitative aspects. The qualitative element, phase one, consisted of personal interviews with proprietors and/or managers of all hotels in the area (seventeen), gift shops (four), and tour operators (three). The purpose of the qualitative assessment was to understand basic business and management practices. Subsequently the quantitative work was completed and consisted of a review of all bookkeeping records of the businesses interviewed in phase one. The data collected in this phase were analysed by computing standard management and operating ratios. The intent of this work was to determine if there were any quantitative measures of performance which could be utilized to understand business decisions and help with organization of training programmes. The information that is described below deals only with that for the hotel sector.

CHARACTERISTICS OF THE HOTEL SECTOR

The development project described above actually began implementation in 1991. An inventory of tourist services preceded implementation and revealed that the total number of international standard rooms in the area was less than ten. A number of small budget hotels also existed, but they catered primarily to funeral attendees. It is common in Ghana for people to be honoured by a funeral some time after they have died. The body may be buried immediately after death or, if the family has the resources, put in cold storage, but the funeral will be held at a later date regardless of the disposition of the body. When a funeral date is decided it will be announced for some time before the event. Posters stating the dates for the funeral and containing a short biography of the deceased are printed and distributed. Funerals generally begin on Saturday and depending on the status of the deceased may continue throughout the weekend. It is expected that anyone who had dealings with the deceased, came from the same village, or was a friend of the family would attend or send a representative. Because this would entail at times long-distance travel for a prolonged event, a budget hotel industry developed to cater to the mourners. This was the extent of Central Region tourism centred around Cape Coast and Elmina in 1991 when the project began.

Table 10.1 provides a good picture of the development of the hotel sector. Notice that in 1995 a record number of new hotels opened. To the casual observer this might seem to coincide with the development at Kakum Park, in particular the opening of the canopy walkway, and the inauguration of a new museum in Cape Coast Castle. However, other information reveals a different story. In 1994 the government of Ghana passed a new investment code. Contained in that code were special provisions affecting the hotel sector. The two most important were a tax reduction from 35 to 25 per cent of net profits from hotel operations and duty-free importation of hotel equipment or furnishings, for example air conditioners, tables and chairs (Ghana Investment Promotion Centre Act 1994). A further incentive for hotel development came from the provision of low-interest loans (referred to in Table 10.2 as SSNIT and CEDECOM loans) backed by the Ghana Tourist Board. These loans are from bank sources and as such may be considered to be part of the formal loan sector which is expected to serve a much different function than an informal sector loan (Alila 1996). This line of thinking will be explored in more detail in the following discussion section.

What is done with year-end operating profits? For most of the businesses there are no year-end operating profits which they must allocate. Most of the businesses do implicitly reveal they have earned a profit. However, they prefer to reinvest or, as stated in Table 10.3, 'plough it back into the business'. This should not be an unrealistic finding for relatively new businesses which are in the development stage but observational data indicates that only five of those indicating they reinvest the profits in the

Table 10.1 In what year was this business started?

Year business started	*Number of hotel/restaurants*
Before 1970	1
1970–1979	3
1980–1989	1
1990	1
1991	1
1992	1
1993	0
1994	1
1995	6

Source: Bowditch 1996.

Table 10.2 What is your attitude towards loans?

Attitude toward loans	*Number of hotel/restaurants*
Don't have any loans	3
Don't want any now – interest rate too high	2
Have SSNIT[1] loan	7
Have CEDECOM[2] loan	2
Will take if rates reasonable and can pay back	3
Can't get any/No collateral	1

Source: Bowditch 1996.

[1] SSNIT=Social Security Service.
[2] CEDECOM=Central Region Development Commission.

Table 10.3 What do you do with year-end profits?

Use of year-end profit	*Number of hotel/restaurants*
Don't have any yet	3
Plough it back into the business	11
Take out a little each month	3

Source: Bowditch 1996.

business actually show, at least to the casual observer, some developmental change.

Part of the answer to what happens with profits may be revealed in Table 10.4 which shows that ten of the fourteen hotels responding hire friends or relatives of the owner. This is in keeping with the Ghanaian culture of assisting an extended family member and may be one of the reasons for the response that profits are reinvested, as that phrase may also mean the write-off of costs incurred from conducting business.

The final table (Table 10.5) shows the importance of external influences in business decisions. When asked to respond to whether they feel business owners seek input from a priest or priestess before making a business decision the overwhelming majority responded in the positive. Only one indicated this would not happen. It should be mentioned here that a priest or priestess in this context is not an ordained minister but rather someone who is in contact with the spirit world. This does not mean people who seek information from the spirit world deny the existence of a God. Most Ghanaians believe strongly in God but also in spirits who do a great deal more to control everyday life than an all powerful God. Sarpong (1974) explains the role of spirits in daily life and the influence they have on behaviour. When the same hoteliers were asked if they consulted a priest or

Table 10.4 Are hometown friends, family or acquaintances of the owner(s) employed here?

Friends, family, acquaintances employed?	*Number of hotel/restaurants*
Yes	10
No	4

Source: Bowditch 1996.

Table 10.5 Is it true that some business owners see priests or priestesses when making decisions?

Consult priest/priestess in business decisions?	*Number of hotel/restaurants*
Yes	12
Maybe	2
No	1

Source: Bowditch 1996.

priestess before making a decision they all responded in the negative. The role of the spirit world is then bound up with one's culture which, as it appears, may not be easily revealed to outsiders but yet may have a powerful influence on decision making.

The above results all came from the qualitative portion of the research study. As mentioned this was followed by a exhaustive review of each business's financial records which were used to construct management and operating ratios. The prevailing view of the African entrepreneur is of one who works hard but loses money because of poor management and lack of financial records (Kinyanjui 1996). The lack of financial records was not a problem for the hotels studied. However there was a great deal of variation in the ratios, leaving one to question the credibility of the financial records supplied. In spite of this, there were some consistencies with the previous information presented. For example, ratios that relate to profit such the Return on Assets, Gross Margin, and Net Income all revealed that profits were pretty much non-existent for the surveyed properties. This is consistent with the findings that only three hotels indicated they were able to generate a little something each month to take out of the operation.

Debt ratios were also calculated. Until 1994 when the country investment code was modified favouring hotel development and the subsequent provision of low interest loans, very low debt-to-equity ratios were recorded. This changed significantly at the end of 1994 when the low-interest government-backed loans were offered. Some hotels' debt-to-equity ratios increased anywhere from 60 to 350 per cent. Unfortunately, due to the need to keep recent information confidential, no data are available for 1995 on. As 1995 was a period of active hotel development it is likely some of the ratios, in particular those dealing with debt, may have changed significantly.

DISCUSSION

What does all this information tell us about hotels operating in the Central Region tourism development sector? Are they better off now than before the development project? Are they able to adjust to the increasing competition? Are they sustainable? Some additional information may help us answer that question.

A majority of hotels (69 per cent) have in the last few years taken on increased debt. Most of that debt (90 per cent) was obtained from formal credit sources (i.e. financial institutions). This type of borrowing is not common for many SSEs, as formal financial institutions have as a matter of course avoided loans to SSEs citing lack of collateral, high cost of managing numerous small loans, and lack of information about how the sector operates (Aboudha 1996). This fear of lending to SSEs is justified based on numerous recorded incidents of high default rates. In the Central Region research results indicate that 78 per cent of the businesses have defaulted on loan repayment at least one pay period. Currently, the lending institutions are in the process of renegotiating payment schedules with most of the hotels. Three are in serious default and have had their loans referred to attorneys for the Tourist Board which stands as the loan guarantee authority.

It should be mentioned that during the entire period from 1991 to the present, tourist arrivals have shown substantial increases for the region. Are hotels not making money as their statements about profits and debt ratios may indicate, or are there other reasons for this finding? A review of some of the literature on financing SSEs may help answer the question.

There are two forms of credit operating throughout much of Africa. These are simply termed informal and formal. Informal consists of loans received from friends or relatives and some credit associations. The most common type of credit association found is called a Rotating Savings and Credit Association (ROSCA). ROSCAs have different names throughout Africa and may be called *njangi* in Cameroon, *chilimba* in Zambia, and *susu* in Ghana. ROSCAs have their historical roots in West Africa (Delancey 1978, from Alila 1996). They developed later in Kenya when it was a British colony, as formal loans to Africans were outlawed under British rule.

ROSCAs take many different forms but all have in common a few traits. Basically a group of people will form an association and agree to contribute a certain amount of money on a regular basis. At specified periods one member of the group will be able to withdraw their entire contribution plus that of the others. Each person takes their turn and receives the same sum of money.

The sum they receive may be reduced, however, by the salary paid to a non-group member who makes the collections, protects the money, and provides the payout. If all costs of participating in a ROSCA are calculated it becomes clear that it is a saving programme offering a negative interest rate! The value of your payout is reduced by inflation, which can be quite severe depending on when you receive your money, and any service fees paid to the collector. On the surface it would appear that participating in a ROSCA is irrational economic behaviour which seems even more so when one considers that official banks do exist that offer interest-bearing accounts. However ROSCAs offer many benefits of which the most important is to secure a large sum of money, at one time, which can be used to purchase expensive consumer goods or pay off other debts such as school fees. Since they are institutionalized and socially accepted they have become culturally important.

Throughout large parts of Africa there are unwritten moral requirements about how money is to be spent. In Ghana certain phrases are quite common and are found colourfully written on boats, buses and signboards. In Twi, the most important to this discussion are: *Onipa baako ndidi* (One man no chop), *Woni sika a woni adamfo* (No money no friends), and *Sika ye mogya* (Money is blood). The meaning of 'One man no chop' is essentially that if you possess money it is your duty to spend (chop) it. If one man holds the money, all will suffer, as we all need a little chop to survive (chop has numerous meanings in Ghana including food). The second phrase 'No money no friends' is somewhat self explanatory but at a deeper level means that a wealthy man will possess many friends as he will feel a responsibility to take care of people. Finally the last phrase 'Money is blood' has an unwritten appendage which is 'and must flow to survive'. Simply what it means is people must eat, and moving money around (i.e. chopping it) will allow others to acquire enough to buy a little food and survive. The phrase 'Chop Economics' has become accepted by the development specialists on the project to explain what at times seems to be irrational economic behaviour. This behaviour is part of the traditional social system in parts of Africa. Hyden (1983) refers to the communal or extended family sharing of resources in Africa as the economy of affection.

The cultural connotation of above phrases is quite simple. If you receive money it is your duty to move it quickly. Buying a bag of cement is sufficient. Holding money is not acceptable. Given these cultural requirements it would be in bad form to save money in a bank account with the idea of making a large purchase (e.g. automobile) even if the formal banking system were readily available. Hence the importance of a ROSCA which allows for the acquisition of large amounts of capital for large purchases without violating any cultural prescriptions. The price paid for this cultural subversion is a negative interest rate.

How does 'Chop Economics' relate to the research reported here? Alila (1996) states that money borrowed in the informal sector often pays for consumable goods and is not used for investment. In his study of ROSCA participants in two villages in Kenya, he found that almost one-third of the money received went to pay for food followed by such things as household utensils and furniture. School fees, which can be quite high, were the only type of long-term investment commonly paid through the use of ROSCA proceeds. Similar findings were reported from a study in Cameroon (Delancey 1978, from Alila 1996). Money received from other informal borrowing sources (e.g. relatives) was also used to purchase consumable goods. It appears, from observation and unstructured interviews with local hoteliers, that money received from SSNIT and CEDECOM loans may have been viewed from the informal borrowing perspective rather than the formal one. If this is the case then money received would have most likely gone to purchase consumable goods rather than for business investment purposes which was the intent of the lending programme.

There is a tendency in Africa to view all lending as informal unless some non-traditional controls are introduced. Even with formal lending institutions, when traditional controls are maintained, such as found in Ghana when the area chiefs control the lending co-operative, frequent abuses are encountered and high rates of default are common (Songsore 1992). Aboudha (1996) reports on the types of credit programmes that have been introduced in Kenya with varying degrees of success. The types that appeared to work best included loans accompanied by business training or co-operatives where high default rates could exclude all members of the credit group from receiving further loans. Neither of these controls was imposed for the SSNIT or CEDECOM loans in the Central Region. Based on the results of the qualitative portion of the research conducted, it was determined that very few owner/operator/managers had any formal hotel business training. They simply moved into the business because they had access to a building or they perceived there were going to be opportunities as a result of the development project. Secondly there were no attempts to include the regional Hoteliers Association in the loan approval process. As mentioned, the Ghana Tourist Board guaranteed the loan, which was done without consulting the Hoteliers Association. Each loan recipient was free to spend the money received as they wished which is exactly how an informal credit scheme operates. It should be mentioned that the Ghana Tourist Board did assist each borrower to prepare a business plan and feasibility study which was a prerequisite to receiving a loan. However there was no attempt to make sure the business plan was followed. The preparation of a business plan would be viewed more as ritual required to obtain money rather than a blueprint for how the money was to be spent. Any money received as a loan would be viewed by the borrower as coming from the informal sector unless certain controls were instituted to make sure repayment obligations were viewed in a different context. Thus from a cultural perspective the borrower could spend the money as they wished.

Another indication of the role of culture in business operations can be surmised from employment practices. An extended family group is part of the Ghanaian social system. The evidence for this is borne out by the number of hotel owner/operators stating that they hire relatives to work for them. In some ways the business operates like a commissary for the extended family rather than a legitimate business concern. Part of this is due to the need for developing social, economic, and political networks which an extended family can provide (Pedersen 1997) and another part is to fulfill the obligations under the chop economics or economy of affection system.

Ghanaians are well known for their hospitality. Visitors are always greeted with *Akwaaba* (You are welcome). It is common for members of an extended family to be offered room and board when they visit. No prior notification of the visit is ever expected. If a stranger accompanies the family member, the same gracious hospitality is offered to them (Sarpong 1974). It is not surprising then to see how many businesses employ family members regardless of skill level. It would be considered a cultural affront to do otherwise regardless of its impact on the business. Some of the problems this entails were uncovered during an in-depth interview with one of the hotel managers at the premier international standard hotel in the region. Prior to the hiring of a non-family manager, the hotel was functioning poorly. Family members of both the husband and wife (joint proprietors) were working at the hotel. The leading member of each family group was attempting to lay claim to the manager's position since the joint proprietors were absentee owners. Food, taken from the hotel, was sold by different family members to people in a nearby village. The proceeds of the sale were never returned to the hotel. Obviously the hotel was an important asset to each side of the family so inner battles (in a low key Ghanaian way) were being waged over control. Both owners realized the situation was bad and getting worse. The hotel was constantly losing money and providing poor levels of service to its guests (a distinction is made between poor service, for example long waits for meals, and gracious hospitality). To remedy the situation an outside manager was selected for the facility. This outside manager had control over all aspects of the operation including personnel. By hiring an outside manager the owners essentially absolved themselves of the cultural responsibility to take care of family members, allowing the business to be run professionally.

One other interesting indication of cultural influences on decision making is the reliance on fetish priests and priestesses (gender is irrelevant) for assistance when business decisions are to be made. Sarpong (1974) relates the importance of priests to the Ghanaian culture. Priests are held in high regard within the community. It is not a position that one can lay claim to even though it may be gained through hereditary succession or conferred by the deity the priest represents. All priests apprentice for the position and must be able to win the trust of the people they serve. There will be at least one priest for each recognized deity. Although there is a belief in an all-powerful

God there are also numerous minor deities which control daily life. Individuals may have a favourite god which they will constantly seek advice from through his intermediary, the priest. Even though in the study no one who was interviewed admitted to seeking advice from a priest, the vast majority were quite sure others did. This indicates that the practice remains an important part of daily life albeit one that is not readily shared with those outside the culture. If this is the case then it is quite possible that economic rationale would be a secondary consideration when seeking assistance for a business decision. The two, economic and cultural, are not mutually exclusive. It is entirely possible that one's favourite deity is also one that provides the best business advice. Nevertheless, the importance of cultural influences in the business life of the Ghanaian entrepreneur appears quite strong.

CONCLUSIONS

The research reported in this chapter is part of a much larger tourism development project in the Central Region of Ghana. Concerns about project outcome sustainability have been raised since the beginning of the work. It was felt that much of what would make the work sustainable would be how the private sector developed and consequently provide benefits to the local population. Therefore it became necessary to organize, train, and in many respects assist the development of the private tourism dependent sector. In doing so it was necessary to understand how SSEs, of which the hotel industry in the Central Region definitely is part, operate in Ghana. Very little information was obtained about SSEs in Africa in particular and nothing could be found that was focused on the tourism industry. Therefore it was necessary to piece together what was available and try to relate that information and respective theories to what was found from both a qualitative and quantitative assessment of the hotel sector in the Central Region. Two prevailing theories, neo-classical and dependency, dominate the literature on development in Africa. Both are macro level theories and, while providing some insights into why development throughout Africa has been retarded, they do not provide enough information to understand local development problems. This study attempted to understand local development issues by interjecting cultural conditions into the equation.

The first issue addressed was one of obtaining financial assistance for business development. The role of the formal and informal credit programmes was examined with special attention paid to the functioning of ROSCAs. These credit schemes are uniquely African, with their roots in West Africa. In spite of their negative interest rate feature they remain popular as, it was hypothesized, legitimate means for avoidance of the 'Chop Economics' system. When other, formal, credit schemes are offered without any attempt to differentiate, through training and other controls, from the accepted informal schemes, limited investment of loan funds and consequently high loan default rates should be expected. This appears to be the case in the Central Region when SSNIT and CEDECOM development loans were made available to tourism interests. This finding should not be interpreted as suggesting that formal credit schemes are not needed or cannot result in high repayment rates. Since SSEs make up the majority of businesses in Africa, with Ghana being no exception, there does appear a need to inject much-needed funds into these businesses. However to be effective the influence of culture must be taken into consideration. The use of co-operatives, which essentially is what ROSCAs are, with formal requirements, which ROSCAs do not have or need, is a must to overcome the need to move money as embodied in the 'Chop Economics' system prevalent in the country.

A second, but related issue, dealt with the need to support family and friends through employment opportunities. Again 'Chop Economics' can be seen as a detriment to the efficient functioning of business as often it results in the creation of an extended family welfare net rather than helping establish a functioning production unit. Many hotels, prior to the development of a regional tourism product, were of the budget variety catering to the domestic tourist who most likely was attending a funeral. The hotel manager then did little to improve service quality as these businesses were for the most part established to take advantage of a small but steady market. While inefficiency and poor levels of service may have been the norm for years and have become acceptable to the domestic tourist, the international visitors that were beginning to show up were used to a different class of service. To take advantage of this opportunity new management systems seem to be in order. Success will be measured by how well the new management systems are able to incorporate efficiency and hospitality standards without violating any cultural norms related to family support. As described above, one hotel was able to overcome this problem by hiring a manager not related to the owners and therefore with no cultural obligations to the owners' family members. This is a legitimate solution entirely within acceptable cultural norms. Another method that may prove acceptable, but there is little evidence it has been utilized, is for the local Hoteliers Association to develop a code of business practices that discourages the employment of anyone that does not possess appropriate skill levels for the position they wish to hold. This would ensure that anyone receiving a position is at least qualified to carry out the responsibilities of that position. It is not the case that people are lazy and wish to take advantage of a family situation. Rather, it should be remembered that the most important consideration is first taking care of family and then taking care of business. By the same token, the family member receiving a position has a moral obligation to his family to perform to the best of his/her ability. Often problems arise because employees' job performance is poor simply because they are unaware of how best to do their work. Once they are trained they can become an asset to the owner rather than cultural baggage. Of course appropriate training must be provided under the auspices of the Association for this to become a solution.

Finally it should be recognized that there are probably more aspects of the culture that have an impact on business operation than were uncovered in the research conducted for this chapter. The finding that many people believe others consult a priest to help make decisions but yet deny that they do the same indicates that there are many cultural elements that are either not shared with those outside the culture or are not openly discussed within the culture. Whatever the case there will be an impact on how businesses operate.

How sustainable are these businesses? Throughout the tourism development literature one constantly finds reference to the need to support or maintain locally owned businesses. Jafari (1988) traces this thinking to reactions from uncovering and understanding tourism's negative impacts both economically and socioculturally. He refers to this body of literature, first appearing in significant numbers in the late 1970s and early 1980s, as the Adaptancy platform. Adaptancy refers to strategies to overcome the influences and impacts of mass tourism. Ecotourism is an example of these alternative forms of tourism recommended to overcome some of the negative impacts associated with tourism development. Implicitly recognized in this body of literature is the idea that small, locally owned is good. But one must question whether small can survive in an increasingly competitive environment and still provide a level of service in demand by visitors.

It would seem on the surface that hotels in spite of the obvious problems discussed above are surviving and do provide tourists with a local experience. However, if the projections made in the tourism master plan become reality the country as a whole will experience a 400 per cent increase in arrivals in less than twelve years and the Central Region, because of its proximity to Accra and its core of world-class attractions, will receive a significant share of those visitors. When that happens profit potential increases and market instability diminishes. In other words, the situation becomes ideal for large-scale enterprises. Already discussions have been taking place with multinational providers to assume part ownership and management of a yet to be constructed resort hotel in the region. Unless the existing properties are able to resolve their present management problems and overcome some serious cultural norms of conducting business they may not be able to survive in the face of the new competition, even in a growth economy.

What this research seems to indicate, in spite of its obvious limitations (e.g. small sample size), is that the prevailing theories offered to understand the lack of development success in Africa may not adequately cover the full range of development obstacles. 'The classical development theories do not offer satisfactory explanations of the continued importance of small and medium-sized enterprises in both industrialized and developing countries' (McCormick and Pedersen 1996). In fact the basic assumption underlying both main theories is that economic factors control development. It is possible that economic factors are of a secondary concern and that socio-cultural obligations are driving decisions. If so then adjustments can be made to western models, such as described in this chapter, for formal credit schemes (i.e. co-operatives and training), to overcome the cultural barriers preventing large scale development of SSEs. This is obviously a bold statement given the limited research conducted for this paper and the wealth of information supporting the neo-classical and dependency theorists. However as Hoddinott (1996) asks 'does an empirical rule developed in the context of western economies fail when applied to Africa?'

There can be no denying the influence of culture on the business management practices of the hotels operating in the Cape Coast/Elmina sister cities areas of Ghana's Central Region. Both the qualitative and quantitative examination of the hotels in the area support this statement. If these results and the hypothesis derived in this paper have validity, further work, on a much larger scale, will have to be conducted. For now the addition of a cultural element to either or both of the development theories would be premature. However if the recognition of the role of culture, and its related social obligations, can aid in the development of Africa's SSE sector, it is research worth pursuing.

References

Aboudha, C. (1996) 'Small scale industrial financing in Kenya: case for venture capital', pp. 193–209 in D. McCormick and P. Pedersen (eds) *Small Enterprises: Flexibility and Networking in an African Context*, Nairobi: Longhorn, Kenya.

Alila, P. (1996) 'Informal credit and rural sector enterprise development', pp. 175–92, in D. McCormick and P. Pedersen (eds) *Small Enterprises: Flexibility and Networking in an African Context*, Nairobi: Longhorn, Kenya.

Anunobi, F. (1994) *International Dimensions of African Political Economy: Trends, Challenges, and Realities*, Lanham, M.D.: University Press of America.

Bowditch, N. (1996) *Micro-Economic Assessment of the Tourism Related Businesses in the Central Region*, prepared under contract to the Tourism Center, University of Minnesota.

Delancey, M. (1978) *Savings and Credit Institutions in Rural West Africa*, East Lansing: African Studies Center, Michigan State University.

Ghana Investment Promotion Center Act, 1994, Act 478, Ghana Investment Promotion Center, Accra: Ghana.

Hansom, D. (1992) *Small Industry Development in Africa—Lessons from Sudan* Lit Verlag: Münster.

Hoddinott, J. (1996) 'Wages and unemployment in an urban African labour market', *The Economic Journal* 106 (November): 1610–26.

Hyden, G. (1983) *No Shortcuts to Progress: African Development Management in Perspective*, Heineman: London.

Jafari, J. (1988) *Retrospective and Prospective Views on Tourism as a Field of Study*. Paper presented at the 1988 Meeting of the Academy of Leisure Sciences, Indianapolis, Indiana.

Kinyanjui, M. (1996) 'Small and medium manufacturing enterprise formation and development in central Kenya', pp. 143–58, in D. McCormick and P. Pedersen (eds) *Small Enterprises: Flexibility and Networking in an African Context*, Nairobi: Longhorn, Kenya.

Liedholm, C. and Mead, D. (1987) *Small Scale Industries in Developing Countries: Empirical Evidence and Policy Implications*, East Lansing: Department of Agricultural Economics, Michigan State University.

McCormick, D. (1992) *Why Small Firms Stay Small: Risk and Growth in Nairobi's Small-Scale Manufacturing*, Nairobi: Institute for Development Studies, University of Nairobi.

McCormick, D. and Pedersen, P. (1996) *Small Enterprises: Flexibility and Networking in an African Context*, III, Nairobi: Longhorn Kenya.

Ministry of Tourism, Ghana (1996a) *National Tourism Development Plan for Ghana 1996–2010*, prepared by the Ministry of Tourism, Ghana; United Nations Development Programme; World Tourism Organization.

——(1996b) *Diary Survey of International Visitors to Ghana*, prepared for the Midwestern Universities Consortium for International Activities (MUCIA)

Ongile, G. and McCormick, D. (1996) 'Barriers to small firm growth: evidence from Nairobi's garment industry', pp. 40–62 in D. McCormick and P. Pedersen (eds) *Small Enterprises: Flexibility and Networking in an African Context*, Nairobi: Longhorn, Kenya.

Pedersen, P. (1997) *Small African Towns–between Rural Networks and Urban Hierarchies*, Aldershot: Avebury.

Sarpong, P. (1974) *Ghana in Retrospect: Some Aspects of Ghanaian Culture*, Accra: Ghana Publishing Corporation.

Songsore, J. (1992) 'The co-operative credit union movement in north-western Ghana: development agent or agent of incorporation?', pp. 82–101 in Fraser and Mackenzie (eds) *Development from Within: Survival in Rural Africa*, London: Routledge.

Teye, V. (1988), 'Coups d'état and African tourism: a study of Ghana', *Annals of Tourism Research* 15: 329–56.

11

SUSTAINABLE TOURISM DEVELOPMENT IN SOUTH AMERICA

The case of Patagonia, Argentina

Regina Schlüter

INTRODUCTION

South American countries, in implementing development plans and programmes since the 1950s, have sought not only to secure economic growth but also to improve the social living standards of their inhabitants with regard to health, education and higher levels of personal income. Such objectives have not yet been fully met. More recently, tourism has been seen as a means of contributing to their attainment. In this context, the introduction of the concept of sustainable tourism immediately met with acceptance, even though the meaning of this term has not yet been fully understood. Sustainable development is interpreted by the World Commission on Environment and Development (WCED 1987) to be a process which enables development without destroying the resources that make such development possible. When applied to tourism, as Godfrey (1996) points out, the concept of sustainable development relates to a form of environmentally friendly tourism. In South America it is ecotourism and adventure tourism that are generally considered to be sustainable.

This chapter begins with a short reference to some current issues related to sustainable tourism. Then a brief description of the patterns of tourism development found in South America during the latter half of the twentieth century will be presented. They will be compared with the model implemented in Patagonia, Argentina, in the 1930s, which markedly resembles the sustainable development model currently in existence. Reference is then made to on-going tourism projects in desert areas of Patagonia, notably to fauna reserves on the Atlantic coast where penguins and a variety of marine mammals are under protection, and to heritage tourism related to Welsh settlement in the region. These projects are examined in terms of the present concept of sustainable development.

SUSTAINABLE TOURISM: SOME CURRENT ISSUES

Environmental awareness does not only concern industrialized countries (Ayala 1995). Developing countries are also interested in achieving economic growth in order to increase the welfare of the population while causing the least possible damage to the environment. However, as Fyall and Garrod (1997) observe, even though the underlying concept of 'sustainability' has been established by academics and the industry has shown great interest in its implementation, it is now being continuously redefined, with few guidelines as to its practical implementation. In developed countries the discussion about sustainability is generally focused on environmental problems. In countries using tourism as an economic growth and social welfare strategy, however, the emphasis is laid on development (César-Dachary 1996). Both Hunter (1995) and McKercher (1993) note that the concept of sustainable development comprises two different aspects; conservation advocates emphasize the non-modification of the environment, while the tourism industry takes up a position closer to development.

As in the case of the tourism industry, developing countries seek to accelerate their growth while preserving the environment and producing goods for a non-traditional market. According to Hughes (1995) a new type of consumer has emerged, one that is complementary to the traditional segment of the hedonist consumer. These new consumers are interested in learning from and communicating with the local population. Hawkins (1994) adds that ecotourism, which apart from favouring closer contact with nature also favours fluid host–guest interaction, is attracting an ever-growing number of enthusiasts the world over.

Owen, Witt and Gammon (1993) maintain that sustainable development does not necessarily conflict with economic growth and that economic vitality is essential to combat poverty, improve standards of living and boost environmental protection. Such a view is shared by governments in South America who, like Butcher (1997), are also aware that sustainable development is not a 'panacea' to fight poverty. In South America it is also recognized that meeting the objectives of sustainable development requires participation by the community. This became more evident from the start of the 1980s when democratic governments came back to power and municipalities started to play an important role.

Although, as pointed out by Murphy (1985), community participation gives each tourism destination its distinctive aspect, such participation is often difficult to achieve. One of the reasons for this is that, as noted by Joppe (1996), municipalities and communities are not always one and the same thing. It often happens that communities cluster beyond municipal boundaries or that a municipality comprises more than one community. To this must be added the inflow of immigrants in search of the great spectrum of opportunities derived from the implementation of new economic activities in a given area. Conflicting situations that arise between 'new-comers' and the 'native-born' can be considerable (Taylor 1995: 488). This is precisely the case in some places in southern Argentina where there has been rejection by the NYC (short for 'born and raised' in Spanish) of the families which arrived to take advantage of economic opportunities offered by industrial expansion plans and, more recently, tourism.

However, when it comes to major issues, the wide range of opportunities created by tourism makes these differences fade, especially when the national or provincial authorities induce the different districts to develop nature-based products or new technologies having a low environmental impact, such as alternative energies.

This influence by the State has always been strong in South American countries, and it has been especially noticeable in the field of tourism. At present, it can also be seen in the adoption of 'sustainability' as a development model. This model is a response to various factors: to the growing awareness regarding the environment by people in various parts of the world; to what are described as 'exogenous market pressures' to develop ecotourism (Boyd and Butler 1996: 557); and to the funds provided – among other institutions – by intergovernmental bodies such as the Global Environmental Fund (GEF) and the World Bank for the development of the advancement of the so-called 'soft technologies'.

TOURISM DEVELOPMENT IN LATIN AMERICA: A BRIEF OVERVIEW

Tourism development is usually related to prevailing economic models in Latin American countries which have given rise from time to time to prescribed activities usually set out as examples to be followed (Lizama 1991). During the first decades of the twentieth century European tourism development was viewed in Latin America as a model to be copied. After 1958 the economic model in force was the one developed by the Economic Commission for Latin America (CEPAL). According to this model, Latin American countries should make every possible effort in order to move from the production of primary goods into the industrialization stage. In order to achieve this, it became necessary to develop both the industry and the services sectors, an issue which demanded significant investment and prompted the state to play a predominant role (Schlüter 1991).

With the exception of Mexico and the Dominican Republic, where the government played an active role by investing in the development of integrated tourism centres, not too much of an effort was being made by other Latin American countries. This was not for want of interest, but rather for lack of assurance regarding the profitability of investment: Latin America as a whole was unable to offer the product (sun, sand, sea plus high temperatures) which were consumed en masse by tourists from the United States of America, the largest and nearest tourism-generating country (Schlüter 1994a).

The South American countries which have been able to compete in the international market by providing a sun, sand and sea product are Brazil, Venezuela and Colombia. However the traffic to Brazil suffered from the effects of an international mass-media that stressed negative socio-economic and political aspects which spoiled the image of the country (Schlüter 1993). Due to domestic problems, Colombia and Venezuela, the only two countries which have been able to take advantage of tourism flows towards the Caribbean, began to compete in this market rather late, when other destinations in the area had already been well established.

The economic model propounded by the Economic Commission for Latin America did not come up with the desired results. On the contrary, State intervention in the economy gave way to huge debts with the International Monetary Fund and with the private banking sector. As a consequence, a serious crisis broke out towards the end of the 1980s which called for major economic restructuring. Steps taken included the privatization of State-owned companies, even those critical to national security such as transportation, energy and telecommunications.

At the same time, significant changes were taking place in developed countries as the share of the market that looked for undisturbed nature as its vacation destination increased in the major tourism generating countries (Schlüter 1994b). Developing countries began to express concern over the deterioration of the environment although the sun, sand and sea product retained its importance. South American countries tried to benefit from this new ecologically-minded trend of tourism and devoted much effort to adapting to this demand by creating ecotourism products. Although the Galapagos Islands (Ecuador) were already a significant destination for nature-seeking tourists, countries within the Amazon basin, as well as areas in Patagonia (both Chilean and Argentine), were the ones to benefit most from this new trend.

Despite the creation of products for the long-haul market, inter-regional tourism continued to rank as the most important (64.4 per cent in 1995). However, World Tourism Organization (WTO) statistics show an average annual growth rate of 7.1 per cent during the 1986–96 period of tourism originating in Europe; of 6.75 per cent stemming from East Asia/Pacific; and of 6.1 per cent arriving from North America (WTO 1997). Nevertheless, South America as a whole only accounted for 12.3 per cent of the total number of arrivals to the American continent and for 10.7 per cent of tourism receipts. As can be seen from Figure 11.1 Argentina, Brazil and Uruguay are the main destinations in the region. As regards tourism receipts (Figure 11.2), Argentina, Brazil and Chile are the countries that have performed best.

Destinations with the greatest growth rates in arrivals during the period 1986–96 were Bolivia (10.9 per cent), Chile (9.6 per cent) and Argentina (9.2 per cent). The average annual rate of increase of international tourism receipts during the same period was 17.1 per cent for Chile, 16.4 per cent for Bolivia, and 15 per cent for Argentina (WTO 1997). The growth of

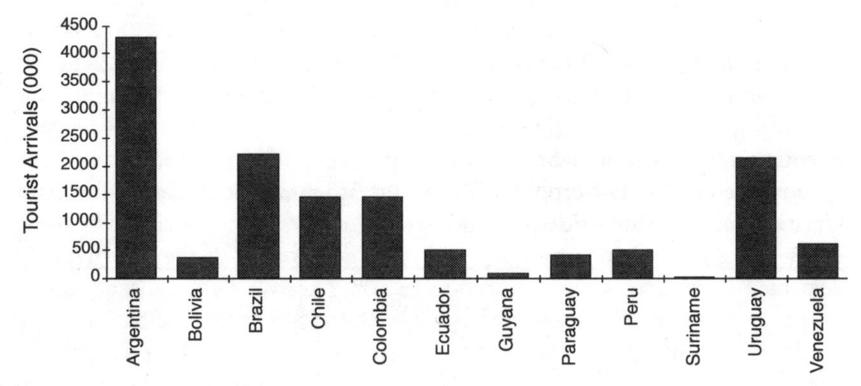

Figure 11.1 South America: tourist arrivals, 1996.
Source: WTO.

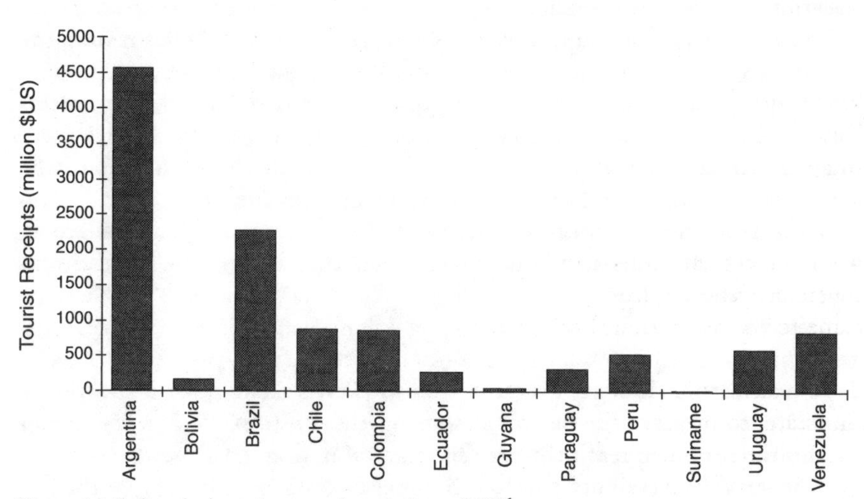

Figure 11.2 South America: tourist receipts, 1996.
Source: WTO.

international tourism in Bolivia exceeded average regional figures. This was due to the creation of new tourism products that fitted into what ecotourism is supposed to be.

Insofar as Argentina is concerned, preference for nature-based tourism contributed to a consolidation of tourism in Patagonia, a region which attracts about one-third of the country's international tourism according to figures provided by the national tourist organization. The region's most visited destination is the city of San Carlos de Bariloche, which is surrounded by the Nahuel Huapi National Park. Its growth is closely linked to the creation in 1934 of Argentina's National Parks Service.

THE ROLE OF THE STATE IN THE DEVELOPMENT OF TOURISM IN ARGENTINA

Although tourism offices became part of some of the provincial administrations, the Federal Government also assumed its role in managing tourism activities on a nation-wide scale by creating the National Park Service. When this department was set up in 1934, two protected areas had already been mapped out:

- Iguazu Falls National Park, on the north-west frontier with Brazil, and
- Nahuel Huapi National Park, in Patagonia, bordering the lake of the same name.

Most of the action taken by the National Park Service tended to centre around the Nahuel Huapi Park because it was more accessible than Iguazu Falls at that time and there was less public interest then in touring forest areas.

The Nahuel Huapi National Park lies at the foot of the Andes mountains (Figure 11.3). Early in the twentieth century it was visited by such well-known personalities as President Theodore Roosevelt of the United States, Edward, Prince of Wales, and members of the Argentine elite. People visiting San Carlos de Bariloche, then a small town, marvelled at the landscape composed of lakes, rivers, woods and snow-capped mountains.

The surrounding countryside was sparsely populated and communications with the rest of Argentina very poor. These features weakened Argentine sovereignty over a frontier area, and officials of the National Parks Service came to the realization that although:

> . . . clearly, our legal function was to preserve nature in its wild state, to preserve the beauty of the scenery and to make it possible for the people to reach the region; it was also necessary to reinforce a theoretical sovereignty that was reluctantly practised and constantly challenged, and that needed urgent support and more solid bases in order to eventually become definitive and unshakable . . .

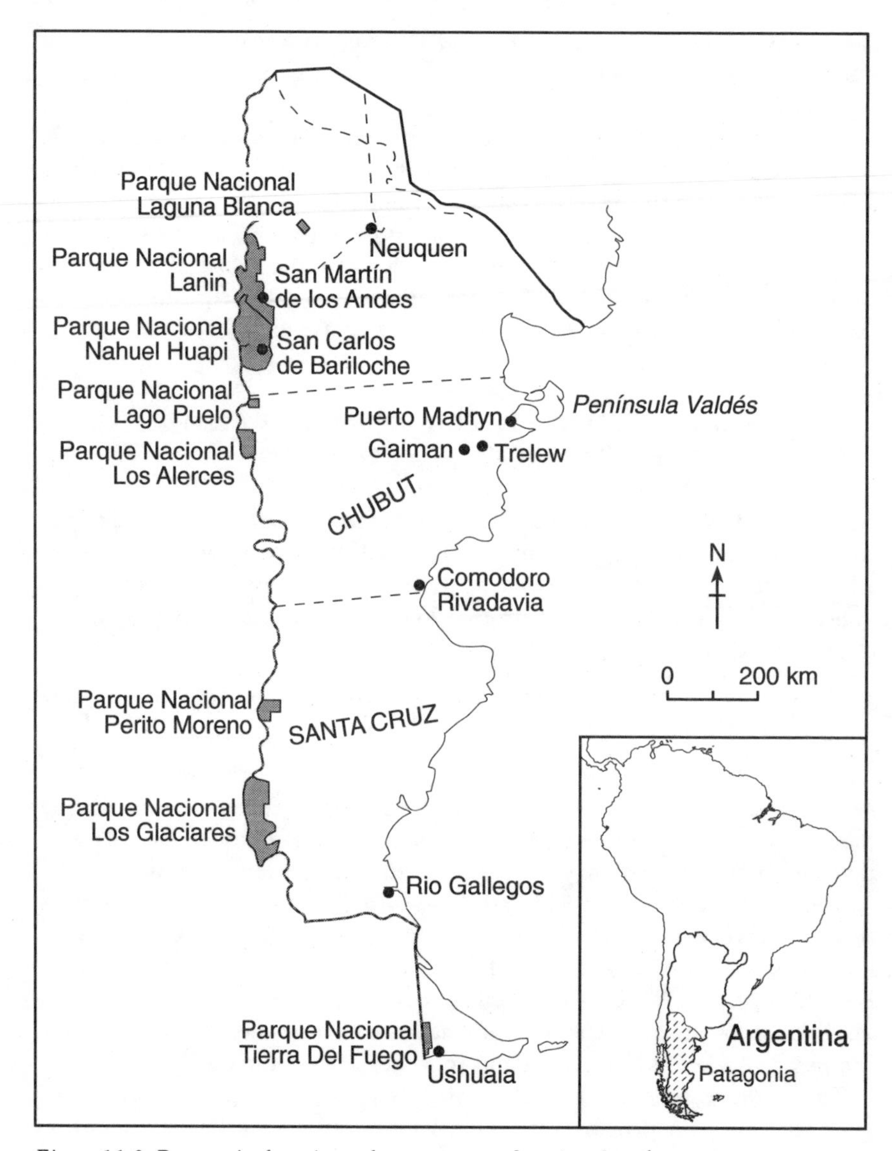

Figure 11.3 Patagonia: location, place names and national parks.

> If Argentina is to be some day, once and for all, the actual owner of the whole territory which, with its lakes, rivers, forest, mountains and ice meets all the destinations of the world, it has to use tourism as an outpost within the framework of the rational preservation of nature and a carefully planned programme of settlement.
>
> (Bustillo 1971: 14)

In order to turn San Carlos de Bariloche into a destination for international tourism, construction of the railway joining it to Buenos Aires was completed. Hotels were reconditioned, sanitary services (sewerage and drinking water) improved, architectural features enhanced and recreational facilities, including ski runs, constructed. The Llao Llao hotel was built to accommodate sophisticated, high-income tourists. Construction and other work carried out pursuant to National Park Service policies directed at fulfilling tourism requirements created a strong demand for outside labour. No fewer than 200 workers were employed prior to the inauguration of the Llao Llao hotel. Once completed, the workers stayed on and switched to tourism-related activities; the slow transformation from an agricultural and stock-raising economy to one oriented to the service sector had already begun.

All these improvements influenced the number of arrivals, which increased considerably. In the 1937–38 season 2,560 tourists arrived. By 1942–43 this figure had risen to 10,900 (Eriksen 1970). The number of visitors from abroad, who came mainly from the neighbouring countries and from the United States, began to grow in the 1960s. The total number of arrivals increased from 45,400 in 1960 to 151,000 in 1967. By the late 1990s San Carlos de Bariloche was visited by 680,000 tourists a year.

The population also grew at a fast rate. In 1915 there were only 1,000 inhabitants; by 1930 the figure had doubled. During the second half of the century San Carlos de Bariloche was the city that had experienced the largest population growth in between two censuses taken in the Patagonia region. In 1960 the city had 21,960 inhabitants; in 1970 this figure rose to 26,799 inhabitants; in 1980 to 48,224, and in 1998 the city totalled 81,130 inhabitants.

As regards demographic structure, in the late 1960s there was a high level of female participation in the economically active population, and a more even gender balance. There was continuous migration of women from ranches in the plateau towards San Carlos de Bariloche. The process started in the stately homes of large ranches, where young women worked as maids and had learned the skills that would later enable them to work in the hotel and restaurant industry (Landoni 1988).

The entire economic activity of the city centred on catering to visitors. The predominance of the tertiary sector of the economy over the primary one made for a clear distinction between this city and the rest of the urban centres in Patagonia. The farms around the city began to produce food supplies such as milk, cheese, butter, eggs and fruit. These supplies were then sold directly to the tourists, especially to those camping out. Small landowners frequently turned their facilities into restaurants and boarding houses. In most cases, however, they subdivided and sold part of their land in order to build tourist accommodation and camping sites. Along with the development of the tertiary sector of the economy, a productive souvenir industry was launched.

This industry produced items such as home-made chocolate, jams, pottery, wall carpets, hand-knitted fabrics, carved wood souvenirs and cosmetics.

In the early 1970s the mountain area close to San Carlos de Bariloche became an attraction for many young people residing in the large urban centres in Argentina, particularly Buenos Aires, who had already visited the area and now wanted to live closer to nature. They devoted their efforts to serving tourists as well as growing berries and aromatic plants, then little known and appreciated in Argentine markets.

Over a ten-year period, and because of their acceptance among tourists, products such as jams, marmalades and preserves secured a privileged position in the main consumer areas and sold as 'souvenirs' or Patagonian 'regional products'. They have gained export status, thus strengthening the regional economy. The mountain range area between parallels 39 and 43 degrees South has become the uppermost producer of berries and aromatic herbs in Argentina. In contrast to what Jurdao Arrones (1992) observed in Spain, tourism provided the basis for agricultural production rather than destroying it.

Although there were some negative effects of tourism development in San Carlos de Bariloche, the balance turned out to be positive. As a result other Patagonian provinces strengthened their economies by establishing their own protected areas. An example of this is the Chubut province, which developed a system of marine fauna reserves on its desert coast.

REDISCOVERING PATAGONIA: FAUNA RESERVES OF THE ATLANTIC COAST

The province of Chubut established fauna reserves on its Atlantic seaboard to promote economic development through tourism based on visits to protected areas. Unlike the case of San Carlos de Bariloche, where tourism was the factor generating basic infrastructure and services, fauna reserves in Chubut benefited from the existence of an aluminium processing plant at Puerto Madryn (Figure 11.3). Whales, penguins, sea elephants and sea lions were the main fauna that attracted people. These species suffered from vigorous commercial exploitation early in the twentieth century. In the early 1960s they were already protected under both federal and provincial laws but lack of proper implementation made them an easy prey for poachers. In 1964 a governmental agency – Dirección Provincial de Turismo del Chubut – was established. It was to take care of two different issues: protection of an extremely fragile resource and management of an expected large tourist inflow. The first step envisaged was the creation of fauna reserves, the system opted for being one that would be endorsed by a clearly defined protected-area network. This concept had been adopted in 1947 at the Brunnen Conference in Switzerland sponsored by the International Union for Conservation of Nature and Natural Resources (IUCN). The conservation programme drawn up to meet tourism objectives was officially adopted in 1967. It was followed by several provincial laws which gave rise to a variety of conservation and management programmes which governed the use of the protected areas and distinguished between Natural Tourist Reserves and Natural Tourist Areas on the basis of their different management features. The establishing of these reserves was greatly praised at international meetings relating to tourism and protected areas, as well as by inter-governmental bodies such as the IUCN and WTO. Península Valdés and its fauna reserves was promoted the most by publications such as National Geographic and documentary films by Andy Pruna and Jacques Cousteau.

Initially the arrival of tourists exceeded the small hotel capacity at Puerto Madryn and the nearby city of Trelew. The development of both cities and their surrounding areas had first begun as a consequence of the poles set up for promoting fisheries and industrial activity. These development poles were established within the framework of the CEPAL model. The latter consisted of a system of incentives based on tax exemptions and promotional prices designed to favour public utilities and infrastructure. The regional economies began to flourish due to the multiplier effects of centralization and specialization, thus attracting a large number of workers both from surrounding rural areas and from more distant regions in Argentina.

Early in the 1990s, significant changes occurred in Argentina affecting political, economic and social conditions. Tourism was then envisaged as a valid option, tending to offset, at least partially, such conditions. An added benefit resulted: that of diverting tourism towards fauna reserves. The main steps adopted were as follows:

- The incentives system was curtailed. All public utilities underwent a process of privatization; subsidies to private enterprises were cut off. This brought about the closure of most industries lying in the Patagonian coastal area, consequently affecting employment. Workers then unemployed tried to switch to the tourism sector.
- Respect for the environment increased, and an improvement in the general behaviour of people was noted. There now emerged a new kind of domestic traveller, keener on being in closer contact with nature and on altering it as little as possible. The tourist industry adapted to these new requirements and took into consideration various factors relevant to preserving the environment.
- The creation of an image of the 'macro' product consisting of whales, marine elephants, penguins and sea lions was worked out by the National Tourism Organization covering the entire length of Patagonia's Atlantic coast.
- The municipal authorities not only recognized the importance of tourist-generated income but also became aware of potential negative

impacts due to the absence of research on the impact caused by human presence vis-a-vis the marine fauna and by the disorderly growth of coastal cities resulting from industrial development and fisheries. The United Nations, the New York Zoological Society and the GEF, together with one of the NGO's offices stationed at Puerto Madryn, surveyed prevailing conditions in the coastal ecosystem and the impact on it of diverse economic activities. At present, both federal and state organizations acting jointly with those listed above are in the process of jointly planning for the management of the Patagonian coastline.

WELSH SETTLEMENTS AND TOURISM

The first Welsh settlers in Patagonia landed on the shores of the present city of Puerto Madryn in July 1865. Whales and sea lions watched their arrival. Due to the scarcity of water there, they went sixty kilometres southwards and settled definitively in the Chubut river valley. Agriculture was their main activity. An artificial irrigation system was built and the groundwork laid for what were to become two important cities in Patagonia: Puerto Madryn and Trelew. While strongly adhering to their native cultural life, particularly religion and language, they also integrated with Argentine social and political life.

The town of Gaiman – approximately 2,300 inhabitants – has become the Welsh cultural centre in the Chubut river valley. When tourists arrived there attracted by the coastal fauna, they also became interested in diverse aspects of Welsh life. Some of the resident families seized on the opportunity of opening their houses in order to serve tea with home made cakes, pastry, bread, butter, marmalade and cheese. At least for Argentinians, unaccustomed to the usual 'five of clock', the 'Welsh tea' was a perfect novelty. 'Welsh tea' was rapidly included in the tourist product of this area. In addition, the Eisteddfod – a poetic and literary contest held each year – brings groups of Welsh visitors from various countries.

Faced with an increasing influx of visitors, the Welsh community joined forces with municipal officials in order to secure greater status in the tourist market for their cultural heritage. An incentives system was established comprising municipal tax exemption for all new building in keeping with Welsh architectural style. Some historic buildings were refurbished or recycled. Various tours were designed, allowing for an appreciation of everything Welsh in Gaiman, ranging from the chapels built by the first settlers to the landscape resulting from artificial irrigation in a total desert environment to old-fashioned wheat mills that stand as witness to an age in this area that was mainly wheat producing. It is mainly the members of the Welsh community proper who undertake responsibility for interpreting their heritage to visitors but it is the municipal authorities (mainly of Welsh descent) who lay down the statutes addressing heritage preservation and its exposure to visitors.

If this Welsh heritage is viewed within its broader Argentine context, an interesting point arises: whose heritage is this? If, as pointed out by Ashworth and Larkham (1994), heritage implies a selection and interpretation of the past, what is the weight that the Welsh heritage carries within the framework of Argentine history? Such an issue also applies to Welsh visitors from abroad, whose provenance is far from what Carr (1994: 64) described as a 'land of castles'. It is probably because of this that, over and above it being just a heritage site, the Welsh settlement in the Chubut valley serves as an outstanding example for both Argentine and Welsh visitors of the ability of two different cultures to defy and challenge nature with only rudimentary tools, and yet build a small and picturesque village on a stretch of barren desert land, drawing from the latter all the necessary elements with which to establish and foster a rich cultural life.

TOURISM IN PATAGONIA FACING THE THIRD MILLENNIUM

Fifty years after tourism was launched in San Carlos de Bariloche that city continues to be the foremost tourist destination in Patagonia (Figure 11.4). Furthermore, it has become the international image of tourism, not only for the surrounding region but for Argentina itself. This is well demonstrated by it being the venue for recent meetings of such top-ranking tourist associations as the AIEST (Association Internationale des Experts Scientifiques du Tourisme) and regional meetings of the WTO. In addition, many other important organizations hold their annual meetings there. It is also one of the sightseeing or rest items included in the agenda of visiting foreign officials as, for instance United States President Bill Clinton during his 1997 visit to South America.

The most attractive places in the Patagonia Andean range come under the jurisdiction of the national parks system. Tourist destinations other than San

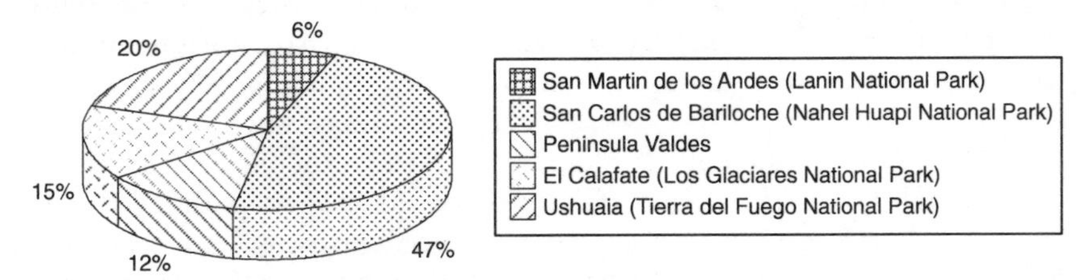

Figure 11.4 Patagonia: main international tourism destinations (per cent of arrivals).

Source: SECTUR 1998.

Carlos de Bariloche gained impetus from the early 1990s when steps were taken in the area to ensure that such growth was closely controlled in order that the diversity of ecosystems did not suffer. Moreover, the 'cordillera' or mountain ranges are traversed by passes which link tourist destinations on the Argentine Patagonia side with similarly valuable locations in Chile. The authorities on both sides work together in developing routes that join the Pacific with the Argentine Andes area, across the desert plateau to the Atlantic Ocean seashore abounding in marine fauna. Promotion of this triple feature has resulted in an increasing flow of visitors from abroad.

In the beginning, tourism was fostered as a means of strengthening and providing support to Argentine sovereignty in the region: paradoxically, it has served to tear down the restrictions of the frontier. At the same time, it has contributed to economic growth and to the conservation of fragile ecosystems.

This constant movement of people from north to south and from east to west made the residents of the Patagonian plateau aware of the economic chances this activity brings about. Therefore, two new forms of tourism are starting to grow in importance: rural and ethnic tourism.

Sheep breeding was Patagonia's main economic activity. Soil damage and problems in wool marketing are bringing ranches to the verge of bankruptcy and creating serious socio-economic problems among the local population. As a complementary source, tourism is to some extent offsetting this lack of income but it also brings with it some problems. There is a need for adequate training programmes and knowledge of marketing techniques in order to reach distant markets. Also, a tremendous lack of infrastructure, especially water, makes it difficult to cater for large numbers of tourists.

With the assistance of NGOs, several projects which pay close attention to the natural and cultural environment, their economic feasibility and the need to provide visitors with a fulfilling experience are currently being developed in the region. A number of action steps, some of which have already been executed and some still in the development stage, may be mentioned: non-conventional water energy production (wind and sun energy), diversification of food production for local consumption, and development of training programmes aimed at the local population so that they learn to make rational use of natural resources and get the necessary skills to cater tourists needs. However, at this stage it is not possible to predict the results.

CONCLUSIONS

Development models that were applied in South America in the past did not meet the varying needs of each of the countries involved. Rather, they had been patterned by intergovernmental organizations on models then prevailing in industrialized countries. Initially, in the 1950s, the first development model enforced disregarded the environment; on the contrary, it aimed at exploiting available resources in order to achieve industrialization. The environment was viewed only as a collection of goods of instrumental value for people. Later on, but again due to external forces, policies underwent a change in the opposite direction, producing a development model primarily interested in the conservation of the environment.

In Patagonia, as early as the 1930s, a development strategy guided by intuition was implemented which resembled closely what today is known as 'sustainable development'. The national parks system served as a pretext to settle population in an isolated area using tourism as the region's main economic activity. Moreover, in order to preserve the area without modifying it, guidelines were set.

Addressed at achieving greater protection of the resources without impairing tourist inflow, a new kind of protected area was introduced on the Atlantic coast. An attempt was made to capture a sector of the market that twenty years later would be known as 'ecotourism'. At that time renewable energy resources, handling of solid waste, modern management techniques and studies to determine environmental impact were not available. Consequently with the increasing number of tourist arrivals the environment paid the price of 'pioneering', but losses were not significant.

From the development of tourism in Patagonia some lessons can be learned in order to make tourism 'sustainable'. Perhaps, the most important of all is not to be dazzled by what is now called 'ecotourism'.

Tourists visiting an area to be in close contact with nature and to learn about the region's environment also require services in order to enjoy their stay. Aware of this situation, the local population makes a living by providing services. Once tourism becomes an important activity a new process starts: people from depressed areas start to migrate to the region, placing a new burden on its resources, similar to the one placed by tourists. In the case of fragile ecosystems the consequences are even worse because the disappearance of any of the markers of environmental quality might reduce or stop the tourism flow.

Another important issue is the need to create a legal system, both at the national and provincial level, that enables governmental agencies to set regulations and have strict control on the use of natural resources and on the management of fragile ecosystems of interest for tourism as well.

The role of the community in the development process is a very important issue. Programmes should not only focus on the tourist, but also foster the enhancement of the quality of life of the local population. If residents share profits derived from tourism in a fragile environment, they soon become the best advocates of nature protection.

Finally, the transfer of technology from more developed societies to the local communities is of great importance. Adequate knowledge is needed to avoid resource overload, to provide efficient administration of foreign financial assistance and to involve the local population in the conservation of the environment in order to ensure economic growth and sustainable use of the resource.

References

Ashworth, G.H. and Larkham, P.J. (1994) 'A heritage for Europe. The need, the task, the contribution', pp. 1–9 in Ashworth and Larkham (eds) *Building a New Heritage. Tourism, Culture and Identity in the New Europe*, London: Routledge.

Ayala, H. (1995) 'From quality product to eco-product: will Fiji set a precedent?', *Tourism Management* 16 (1): 39–47.

Boyd, S.W. and R.W. Butler (1996) 'Managing ecotourism: an opportunity spectrum approach', *Tourism Management* 17 (8): 557–66.

Bustillo, E. (1971) *El Despertar de Bariloche*, Buenos Aires: Casa Pardo.

Butcher, J. (1997) 'Sustainable development or development?', pp 27–38 in M.J. Stabler (ed.) *Tourism & Sustainability. Principles to Practice*, New York: CAB International.

Carr, E.A.J. (1994) 'Tourism and heritage: the pressures and challenges of the 1990s', pp. 50–60 in Ashworth and Larkham (eds) *Building a New Heritage. Tourism, Culture and Identity in the New Europe*, London: Routledge.

César-Dachary, A. (1996) 'Desarrollo sustentable, turismo y medio ambiente en el Caribe. Una opción válida?', *Estudios y Perspectivas en Turismo* 5 (1): 18–51.

Eriksen, W. (1970) *Kolonisation und Tourismus in Ostpatagonien*, Bonn: Ferdinand Dümler Verlag.

Fyall, A. and Garrod, B. (1997) 'Sustainable tourism: towards a methodology for implementing the concept', pp. 51–68 in M.J. Stabler (ed.) *Tourism & Sustainability. Principles to Practice*, New York: CAB International.

Godfrey, K.B. (1996) 'Towards sustainability? Tourism in the Republic of Cyprus', pp. 58–79 in L.C. Harrison and W. Husbands (eds) *Practising Responsible Tourism*, New York: John Wiley & Sons.

Hawkins, D.E. (1994) 'Ecotourism: opportunities for developing countries', in W. Theobald (ed.) *Global Tourism. The Next Decade*, Oxford: Butterworth–Heinemann.

Hughes, G. (1995) 'The cultural construction of sustainable tourism', *Tourism Management* 16 (1): 49–59.

Hunter, C. (1995) 'Key concepts for tourism and the environment', pp. 52–91 in C. Hunter and H. Green (eds) *Tourism and the Environment. A Sustainable Relationship?* London: Routledge.

Joppe, M. (1996) 'Sustainable community tourism development revisited', *Tourism Management* 17 (7): 475–9.

Jurdao Arrones, F. (1992) *Del Eurofelipismo al Desierto*, Madrid: Endymion.

Landoni, M. (1988) *Patagonia y . . . una Forma Especial de Turismo*, Buenos Aires: CIET.

Lizama, C. (1991) 'Desarrollo turístico de Costa Rica', in CESTUR (ed.) *Desarrollo Turístico de América Latina, Mexico*.

McKercher, B. (1993) 'The unrecognized threat to tourism. Can tourism survive "sustainability?"', *Tourism Management* 15 (2): 131–6.

Murphy, F.E. (1985) *Tourism: A Community Approach*, London: Methuen.

Owen, R. E., Witt, S.F., and Gammon, S. (1993) 'Sustainable tourism development in Wales. From theory to practice', *Tourism Management* 14 (6): 463–74.

Schlüter, R. (1991) *Social and Cultural Impacts of Tourism Plans and Programs in Latin America*, Aix-en-Provence: CHET.

—— (1993) 'Tourism and development in Latin America', *Annals of Tourism Research* 20 (2): 364–7.

—— (1994a) 'The role of the State in the development of tourism in South America', *Téoros* 13 (2): 25–28.

—— (1994b) 'Tourism development: a Latin American perspective', pp. 246–60 in W. Theobald (ed.) *Global Tourism. The Next Decade*, Oxford: Butterworth–Heinemann

SECTUR (1998) Estadísticas de Turismo Internacional, Buenos Aires: Dirección de Estudios de Mercado y Estadísticas.

Taylor, G. (1995) 'The community approach: does it really work?', *Tourism Management* 16 (7): 487–9.

World Commission on Environment and Development (WCED) (1987) *Our Common Future*, Oxford: Oxford University Press.

World Tourism Organization (WTO) (1997) *Tendencias del mercado turístico. Américas 1986–1996*, Madrid.

12

TOURISM AND THE HIMALAYAN TRIBES

Searching for sustainable development options for the Bhotias of the Bhyundar Valley

Tej Vir Singh and Shalini Singh

Tourism and tribology have seldom been seen in a symbiotic relationship – the former has a propensity for eroding fragile resources, the latter invariably demands their conservation and preservation. Despite this apparent contradiction, more and more unspoilt, insular and remote habitats of indigenous people have come under the sway of tourism because of their unique attractions of nature and culture (Price 1996). The rationale behind this questionable tourism development has been the perceived absence of any other option for development in these inhospitable environments (Zinder 1969). Some ethnocentrics argue that appropriate and organic tourism is more sensible than the extraction of mineral resources (Butler and Hinch 1996). Altman (1989) and Parker (1993) in their aboriginal tourism research have tried to establish that economic independence of indigenous people is possible through community-based sustainable tourism strategies which promise to restore, protect and conserve biocultural diversity and community authenticity (Sofield 1993).

This chapter explores the possibilities of introducing tourism as an option for tribal development in the remote Himalayan valleys where tribal people struggle against the harsh environment for their marginal existence. This chapter begins by discussing the Himalayan habitat of these hill tribes and the nature of the ecological and cultural resources. It then provides a brief narrative on the Sherpas of Nepal, who by their prowess and ingenuity have made a success of mountain tourism to enhance their quality of life. Taking a cue from this success story of the Sherpas, there is an effort to examine whether other Himalayan tribes could also practise Himalayan tourism to their advantage. The Bhotias of Bhyundar Valley in the higher reaches of the eastern Garhwal Himalaya (Figure 12.1) have been identified as having potential for the promotion of indigenous tourism after considering their

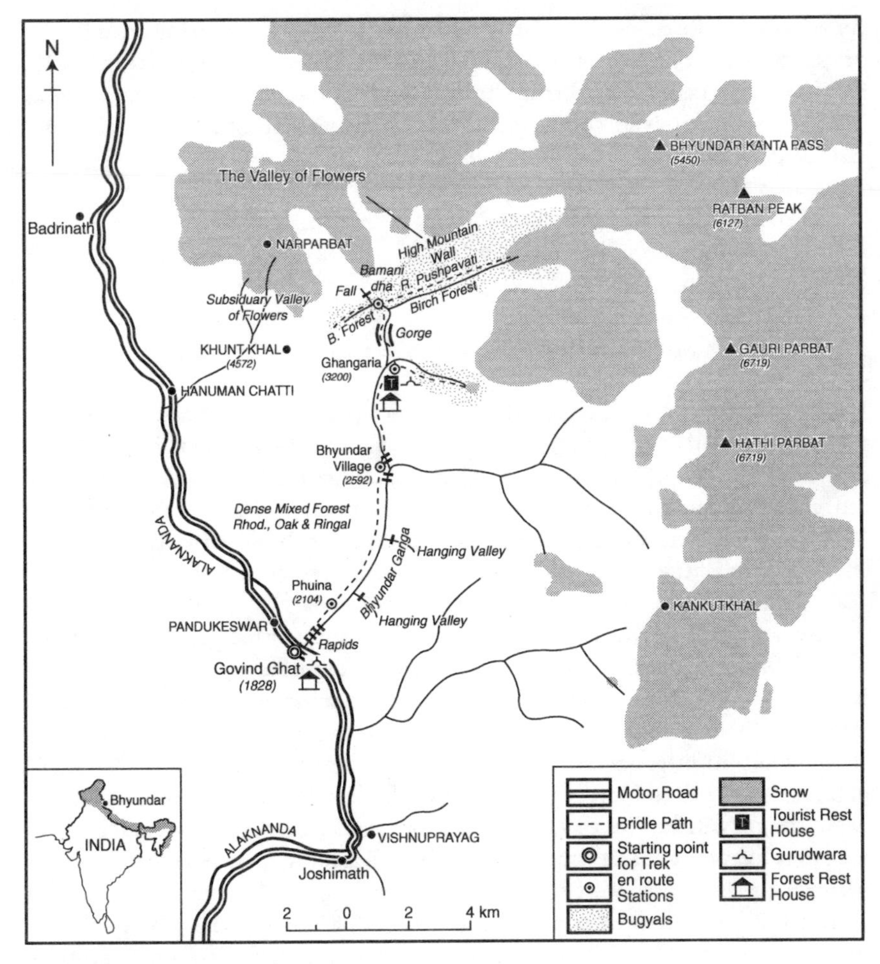

Figure 12.1 The Bhyundar Valley in Garhwal Himalaya.

resource potential and constraints, and the social implications of such involvement.

The main issues to be considered are :

1 the potential opportunities for tourism ventures which exist in and around the Bhyundar Valley;
2 the community capability (entrepreneurial potential) for developing tourism businesses;
3 identifying those options which represent the best opportunities in relation to their acumen;
4 whether the Sherpa tourism model of Nepal could be replicated in this part of Himalaya.

Before taking up these issues, it is pertinent to briefly comment on tourism practices prevalent over the Himalaya. This effort, to some extent, will facilitate the task of selecting 'forms' and 'style' of tourism that these indigenous Himalayan communities can conveniently opt for. Broadly we can identify five practices (Singh 1989):

1 Himalayan pilgrimages
2 amenity tourism of the Raj days
3 park tourism
4 wanderlust tourism
5 tourism in pastoral valleys

Hindu pilgrimages to the Himalayan shrines (*yatras*), with Badrinath as the pre-eminent pilgrim centre, are as old as the Indian civilization and represent India's oldest form of domestic tourism. The practice of this religious tourism was both holistic and humanistic in its holistic approach. It encouraged local benefits, people's participation, the spiritual and social enhancement of the guests and the hosts, besides respect for the environment – all elements being encapsulated in the *Yatra* ritual.

The landscape between the Alaknanda and Bhagirathi, particularly in its upper reaches, is considered highly sacrosanct for pilgrimages, especially the Badrinath zone. The entire Alaknanda valley with all its river junctions (*prayagas*) is a virtual pilgrim's highway, having a floating population of about 400,000 pilgrims and tourists during summer months. It is also the habitat of primitive Marcchayas and Tolcchas of the Bhotia clan, who have been silent witness to the sacrament of *Yatras* being performed by the religious agents (*pandas*). Bhotias by religious sanction were not allowed to perform these services, except peripherally.

The fast pace of the modernization process has altered the whole scheme of religious travel to the detriment of the pilgrimage ethos and the Himalayan society. Many of these religious resorts have assumed a mass tourism character, and present a spectacle of a peculiar mix of the religious and the mundane. Nevertheless with the partial abandonment of the *panda* stronghold, opportunities have been created by the advent of new tourism in which these highlanders can now partake. This aspect will be discussed later in this chapter.

The second phase of tourism development is associated with the British rulers who created 'hill-stations' all along the middle Himalaya for seasonal climatic relief. Thus, the Himalayan resorts of Mussoorie and Nainital had their birth in the first quarter of the nineteenth century and later served the amenity needs of the bourgeois and the barons, rajas and maharajas. After India's independence (1947) they fell prey to poorly planned 'social tourism', facing saturation when 25 million visitors overloaded their capacity. These resorts required specialized services, and the tribal community was ill-equipped to respond to these needs, besides being far away from these tourist foci.

Though India was introduced to the park concept from the time of Vedic Ashramas, park tourism is a recent phenomenon that lays emphasis on three aspects, namely eco-conservation, education and recreation. By and large, it is an elite activity that demands knowledge-based services and adequate park interpretation. In eastern Garhwal there is both India's oldest park (1935), the Corbett National Park in the foothills, and its youngest park (1982), the 'Valley of Flowers', situated in the upper reaches of the Bhyundar, with numerous Bhotia villages en route. In this region Nanda Devi National Park is the largest nature preserve and has a rich biocultural diversity. The park has been closed for visitors since 1982 because of reports of nature-piracy and vandalism. Unfortunately parks in the Himalaya have not been positively linked with local people, and the practice of ecotourism, which should have provided job opportunities to the indigenous people, has not found favour with the foresters and tourism planners.

As the remote Himalayan regions were increasingly opened to tourism in the 1980s, mountain treks, trails and other forms of wanderlust tourism grew popular. Since this form of tourism requires access to some of the most sensitive zones of Himalayan wilderness, environmental controls and proper management are needed to ensure ecological security, better economic benefits and cultural sustainability. Impulsive development coupled with poor planning approaches failed in this regard, leaving the local people disillusioned.

Following wanderlust tourism a new form of tourism has slowly emerged in the distant pastoral valleys, which are blessed with unique community features, a rich rural heritage, religious and cultural attributes and bucolic settings. Since they offer agro-pastoral products, the indigenous communities have a chance to participate in a form of rural tourism. Residents of the Kullu Valley in Himachal Pradesh have taken some advantage of this opportunity, but it is yet to come to the eastern Garhwal. Rural tourism in these Bhotia valleys appears to be desired and to have economic potential, provided it co-exists and does not conflict with the native environment.

The foregoing provides a snapshot of the Himalayan tourism scenario and the potential opportunity spectrum it offers for the indigenous people. In physical setting, it has a dramatic and unique landscape for tourism and in cultural opportunity it has four Hs (habitat, heritage, history, handicrafts) which Valene Smith (1996) recommends as diagnostic tools for the development of indigenous tourism. Despite the fact that most Bhotia villages are located amidst extraordinary Himalayan scenery with comparatively easy access to visitor flows (pilgrims/tourists), the indigenous people have cared little for including tourism in their socio-economic development. While it is true that no attitudinal survey has been conducted, it is quite evident that the pilgrim economy was completely monopolized by the religious agents (*pandas*) and the bureaucracy that runs the tourism industry in Uttar Pradesh. The twin Himalayan development wings (Garhwal and Kumaun Development Corporations) have displayed a meagre imagination for involving local people in tourism. Local residents could have been given better support than that given to entrepreneurs from outside the region. It may be that the planners and developers of Himalayan tourism considered the locals ill-informed, illiterate and unsuitable for the complex activity of tourism. Rural tourism was not yet developed and pilgrimages were considered too austere to sustain an economy. The metamorphosis to secular tourism further complicated the aspects of sustainability. Bewildered Bhotias and Johars of Uttarakhand Himalaya, living a marginal existence, found themselves confused by the economic opportunities presented by the modern forms of tourism.

Having practised Himalayan trade for centuries, the Bhotias in fact have considerable entrepreneurial wisdom for commerce which should be an asset in running tourism businesses. The world knows how their counterparts, the Sherpas of the Nepal Himalaya, have created a niche for themselves in mountain tourism. It is appropriate, therefore, to have a brief discussion on the success story of the Sherpas before answering the question – if the Sherpas can be successful, why not the Bhotias? Alternatively, could the Sherpa tourism model be replicated in the Indian Himalaya?

SHERPA TOURISM

The Sherpas of Khumbu region in Nepal Himalayas present a unique model of indigenous tourism which brought them economic well-being at a time when they suffered a severe set-back to their traditional economy with the closing of the Nepal–Tibet border (1959). The Sherpas are only one of the many Bhotia populations occupying the greater part of Nepal's northern border zones (Haimendorf 1981). They showed exemplary resilience by entering into a new field of enterprise in 'Mountain Tourism' and imparted to it all the skills and spirit of adventure which they had developed as independent traders (Haimendorf 1981, 1984). This new venture offered them more lucrative financial gains than traditional enterprises (Adams 1992). Some of the tourism impact studies carried out in the Everest (Khumbu) region and in the Langtang and Rolwaling Valleys present a fairly optimistic picture, particularly in socio-economic areas, and demonstrate how the development of tourism transformed a subsistence Sherpa society into a cash economy within a short period, raising the standard of living and creating job opportunities in the marginalized areas (Pawson 1984). In one of his earlier studies, Haimendorf (1979) observed: 'Today earnings from employment in the service of mountaineers and tourists have replaced the gains derived from the trade with Tibet, and Sherpas have been able to maintain, and in certain aspects even to improve their standard of living.'

These studies revealed that a Sherpa engaged in trek tourism earned more than the national average (Muqbil 1984). Sacherer (1981) and Baumgartner (1992) explained how a high-altitude Sherpa community in the remote Himalayan valleys, with a narrow economic base, received socio-economic benefits from trekking and mountain expeditions. The Sherpas of Rolwaling were able to manage their local resources, control and regulate their woollen market for porters and also develop other tourist services such as food and accommodation for trekking groups, thereby retaining economic benefits within the valley. Besides, they have been responsible for building better relationships between hosts and guests by working as culture–nature brokers. The Sherpas as caravan leaders have earned a global reputation as the enviable mountain guides and path finders (Gurung *et al.* 1996) and no Himalayan expedition can do without Sherpas, who quite often assume the role of a tour manager (Fantin 1978).

Despite this scenario, it has not been all a rosy picture, for these mountain communities have had to pay the heavy price of environmental damage wherever 'thresholds' were crossed. Garbage trails, litter, alienation and the commoditization of culture were reported in some critical environments (Bjonness 1983; Adams 1992). The most positive aspect of this episode is that these ethnic groups successfully engaged in tourism, studied, discussed and analysed the problems, and developed strategies to capture control over development and decision making for a community-based environment-oriented green tourism (Singh 1996). Campaigns like 'Keep the Himalaya Clean', 'Ecotrekking', 'Polluter Pays' have brought environmental awareness among the Sherpas, which has gone a long way in greening the Himalaya.

The Annapurna Conservation Area Project (ACAP) sets the model for green tourism. This beautiful Sherpa environment initially was invaded by trekkers (36,000 tourists plus the same number of porters) who disturbed the ecological balance of the region, as virgin rhododendron forests were recklessly burnt for tourist energy needs. (There are 850 lodges in the Annapurna Region, and Gurung (1992) estimated that a single lodge in one small village consumes one hectare of rhododendron forest per year to serve the needs of trekkers.) Economic gains from tourism leaked out of the region in the form of imports to meet tourists' demands. ACAP has been able to achieve positive results through the proactive involvement of the local people by creating awareness, training and education, as well as evolving a minimum impact code to be adhered to. ACAP set up visitor information centres for the diffusion of information and feedback and represents an integrated approach to mitigate the negative impacts of tourism on the environment and to preserve the Himalayan heritage. In essence, it addresses three main issues: nature conservation, human development and tourism management (Gurung 1992).

BHOTIAS OF GARHWAL

Ethnographically, Bhotias are descendants of Monkhemer or Kirats (Chatterji 1989; Rustomje and Ramble 1990). They include the Jads, Marcchayas and Tolchhas and Saukas (Kumaun) who inhabit the Bhot (Tibet) region, starting from the southern slopes of the greater Himalayas and extending up to Zanskar, demarcating the Indo–Tibet border. In Garhwal, more of them are to be found in the upper valleys of the Bhagirathi, Vishnu-Ganga and Dhauli-Ganga within a range of 2,250 m to 3,600 m in height, with a highly scattered land area of which only 3 per cent is fit for agriculture. This is, however, compensated for by vast stretches of meadows of Himalayan grasses, herbs and flowering plants, interwoven with junipers and dwarf rhododendrons. These attractive pastures of *mamla* grass (locally *Buggi*) provide highly nutritive grazing for the herds and flocks of goats and sheep, but for over six months of the year these valleys remain snow-bound. Living in this ecological severity, Bhotias have had to establish a pastoral economy based on transhumance with nomadism. Goats, sheep and mules form the backbone of Bhotia lifestyle and handicrafts (Uppadhaya 1988). With two permanent settlements they have their high summer villages and lower winter villages. Summer habitations, located in Dhauli and Vishnu-Ganga Valleys are known as *Maits* (parent village). Winter settlements are known as *Gunsa*. Fourteen *Maits* of the Dhauli Valley have thirty-three *Gunsas* in the Chamoli district. In October these *Maits* are deserted. Almost all Bhotias keep sheep to meet their domestic demand for wool and meat, but not all Bhotias are engaged in pastoral activity, as a flock of less than 200 sheep is not considered an economically viable unit. A flock of 200–300 has to be managed by four shepherds and three Bhotia dogs. Bhotias transhume down the valley in October, and the shearing of wool is done in March and September. At the *Gunsa*, they engage in spinning and weaving.

Bhotias possess marked Mongoloid features. They are fond of ornaments and trinkets and their costumes show a distinct similarity to those of the Tibetans, a trait which becomes more pronounced as the frontier is reached. They are Hindus but Lamaism is vividly manifest in their lifestyle, including the observance of rituals, totemic beliefs and taboos. Despite the impact of modernization, they have preserved their institution of kinship, family and religion. *Ran Bang* is their most colourful festival, when young boys and girls assemble in a house or a field at night to dance, drink and sing. Many young boys scramble on the steep hillsides and collect herbal plants, which helps them prepare for the roles of tourist guides or trail interpreters. Bhotias have an enviable sense of community living, enjoying talking, smoking *Hukka* (pipe) or churning green tea in their bamboo vessel, and are a great attraction for any outsider, specially tourists.

Gaucher, in Chamoli district, is the venue of their Bhot fair where they sell their handicrafts and woollen goods. A Gaucher fair is a reminder of the

Indo–Tibetan trade when Bhotias and other tribes of Mana and Niti gathered together to barter goods. After independence it has been overly modernized and has lost most of its authenticity. This highland tribal community suffered a serious setback after the Sino–Indian clash in 1962 when the Tibetan borders were closed to them. Prior to this event, they flourished on the Indo–Tibetan trade in salt, wool, Borax (imports), Jaggery, and barley (exports). The value of exports and imports maintained a ratio of 1:2 giving 100 per cent profit to the Bhotia at the border (Nityanand and Kumar 1989). Their woollen industry was severely threatened by the political events and they had to find a new economic base in sheep rearing, and thus their pastoral economy became the staple of their economy.

The Bhotia society got a new lease of life through the tribal reservation law that came into force in 1967. With the help of the government they were able to establish their woollen cottage crafts and transport services. The new development measures transformed their way of life, influenced their pastoral economy and encouraged outmigration. During the 1960s about 80 per cent of Bhotia families were engaged in pastoral activity whereas only 20 per cent retained this activity during the 1980s. Few of them are engaged in the pilgrim or tourist economy, which is surprising considering that they live in the most fascinating part of the Himalaya, and are gifted with a trade acumen, hospitable disposition, physical prowess, are deft in crafts and possess an intimate knowledge of the terrain. Such attributes would seem highly suitable for involvement in the tourist trade.

TOURISM DEVELOPMENT OPTION: CHALLENGES

The Sherpa and ACAP models illustrate that indigenous tourism can be an option for the development of tribal societies living in a harsh environment with limited opportunities. In fact no single indigenous model can be a perfect fit, as each unique indigenous culture is constantly evolving in the face of change within the environment in which it exists. Hence diverse and dynamic approaches to indigenous tourism are required (Butler and Hinch 1996). Approaches to development must respect the natural, social and cultural diversity of the area. It should also ensure a pace, scale and size of development (Rodenburg 1989) which protects rather than destroys the tribal culture of highland societies. It should actively support enterprises, co-operatives which provide services and crafts, and encourage 'home-based' tourism accommodation and facilities. This would involve local people in job opportunities and thereby reduce outmigration of young and skilled people. All efforts should be made to promote the unique features of a tribal society without altering them (Eber 1992). There is a need to maintain cultural diversity and authenticity. The profits reaped through these efforts must be channeled back to the area and its inhabitants to foster a sense of pride in them and their tribal and environmental heritage (Niederer 1984). It is best when tribal tourism schemes are integrated with local, regional and national development plans.

Achieving positive results from such a symbiosis is possible only through a redefined development framework – one which is not only economically focused but also based on environmental rationality and indigenous needs. While this would take care of the knowledge-based aspects of the approach to development, there is yet another dimension that must be studied to ensure tourism becomes a successful agent for transforming development. Adu-Febiri (1996) has proposed a 'Human Quality Platform' which has a focus on identifying, explaining and suggesting effective approaches to restructure the hybrid cultures found in many developing countries. Such an approach emphasises those aspects of human capital that focus on dedicated service towards making tourism a development catalyst. This would entail a thorough understanding of the traditions, beliefs, lifestyles and critical human qualities. In the absence of such a detailed examination, it is hoped that a brief investigation of the host community will not only provide a resource database but also suggest appropriate implementation techniques for the identification of development objectives, interest groups, and the type and pace of development, thereby ensuring total community support and involvement. Sofield and Birtles (1996) Indigenous People's Cultural Opportunity Spectrum for Tourism (IPCOST) concept may be helpful in this regard in preparing a more sustainable indigenous tourism plan.

Roughan (1994) who used the IPCOST concept for establishing the Solomon Island Development Trust (SIDT) found that many villages were unable to participate in the development process because of the range of constraints often linked to the structure of a community and its capacity to function as a community. He therefore designed a village self-assessment procedure called a 'Development Wheel' – a key to the self-measurement of village life, its strengths and weaknesses, as related to the development process.

TOWARDS BHOTIA TOURISM: OPPORTUNITIES TO SEIZE

Since indigenous people have, through the ages, evolved the art of co-existence with nature's challenges, it is suggested that they put this knowledge to further use, while tapping tourism as yet another avenue for subsistence. However, one issue that remains implicit is that tourism must not be permitted to become the sole mainstay of the people. It should assume a role in a strategy of diversification that supplements the existing economic structure, integrating into the socio-economic norms, and in a form desired by the community for its social, cultural and economic well-being. This developmental approach suggests that tourism will uphold ethnic authenticity and integrity without causing major deviations from the local lifestyle.

The experience of introducing tourism into fragile environments as economic catalyst has rarely been worth the effort. In the absence of a holistic approach, the tourism function has often failed to deliver the promises of its proponents. Thus, as a rule tourism must be interwoven into the fabric of the human and natural ecosystem. It would be appropriate to recommend a secondary or complementary position/role for tourism, the primary objective being 'sustainable development' of the ecological variables of the 'system'. This may be termed as Development by Objective(s). The 'objective' or purpose of sustainable development should be to realign the values of the stake holders in the right direction.

The next step is to develop an agenda (framework) for the region's development. The agenda must serve as the guiding principles that call for grassroots efforts aiming at insulating the 'indicators' (Hatcher 1996) of the region's biotic and abiotic well-being. These indicators are the indispensable life-forces of the natural and human (co)existence. Hatcher notes five characteristics of a good sustainability indicator:

1 It consistently reflects the status of a significant and fundamental characteristic.
2 It is understood and accepted by the community.
3 It is statistically and practically measurable in logical terms.
4 It has clear, understandable links to other indicators.
5 It represents or directly relates to important community values.

Using these criteria for the identification of key factors as parameters of sustainable development for the Bhyundar Valley habitat and inhabitants, a framework for development can be proposed (Figure 12.2). The region's biosphere, more particularly the floral aspects, is the most vital resource for the co-existence of men and nature. Hence three major ecologically and economically viable proposals – ethnopharmacy, agro-forestry and eco-restoration – are strongly recommended. These suggestions are substantiated with the six Es.

- *Eco-friendly* and nature-based: The health of the biosphere depends on appropriate utilization of biotic resources of species. Forest rehabilitation projects and botanic research projects are suggested with the intention of sound research-based land-use practices for agro-forestry.
- *Ensures* insulation of vital life support systems through the propagation and practice of indigenous farming and innovative recycling techniques. Botanic research must also be aimed at identifying undesirable and potentially destructive species for elimination.

- *Economic* viability of these nature-based functions can be achieved through the application of local resources for the unique and desirable outcomes that can be capitalized upon. For example, eco-restoration projects can trigger a series of functions related to community-based services and resource management. Similarly agro-forestry products can promote the localization of small-scale or cottage industries and indigenous provision centres in the form of wayside amenities. The same could apply to botanic research projects on medicinal plants and species. Economic viability is also reflected in the employment opportunities for the local population that will increase with the introduction of these programmes/projects.
- *Endorsed* by native representation. The Bhotias are a small, closed, high-altitude society with unique traditional mores and practices. They live by the dictates of their environment. Indigenous practices of farming and ethno-botany are endemic life styles which should continue to receive local patronage.

As previously stated, the Bhotias are an attraction in themselves for tourists and also a resource for touristic activities which they service as trek and trail managers, interpreters and guides.

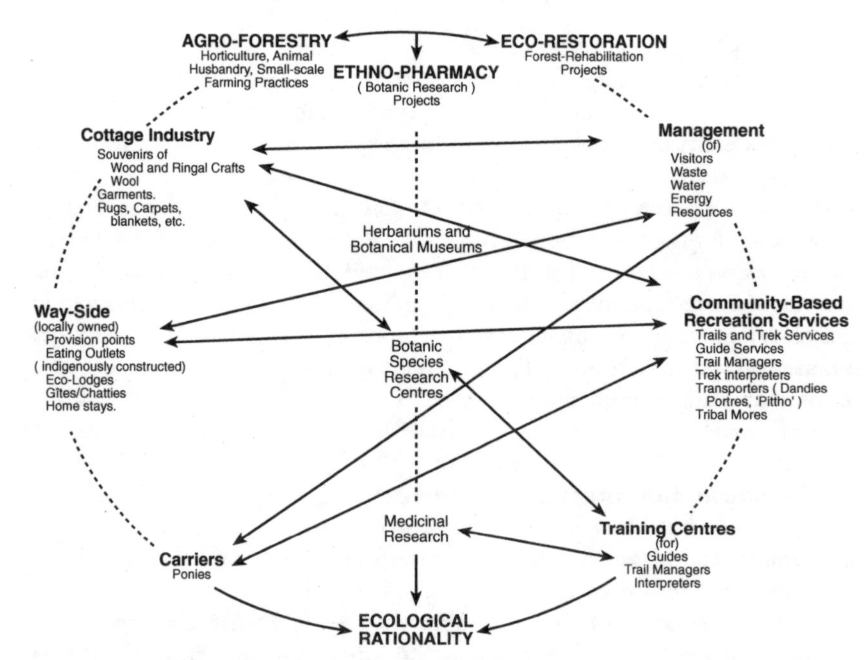

Figure 12.2 A framework for development.

Source: Conceived and Designed by Shalini Singh.

- *Ensures* maintenance and monitoring of the biodiversity through appropriate eco-friendly practices.
- *Encompasses* a broad spectrum and integrated system. By positioning farming, forestry and ethno-pharmacy as the foci of development, a wide range of economic options are made available. Adherence to these guidelines can result in multiple functions that are inter- and intra-dependent in a systematic network.

The aim should be to make planning feasible and beneficial for the people and their habitat, and in this context it would be appropriate to call this form of activity Bhotia Tourism – as this pertains to a site-specific endemic tourism. The term signifies the indigenous nature of tourism which ought to be planned, nurtured and monitored at local/regional levels. This is of particular importance, as rapid and uncontrolled development can destroy the diversity of local, regional structures particularly when they are introduced from outside the region/country. Hence, the term advocates local involvement that promises a strong input of native conservation techniques along with traditional genius (Brandon 1993).

Since community development lies at the heart of such a programme, its prime objective should be to educate and train the involved locals to interact at acceptable levels/limits with the guests, and to cater to their basic requirements without compromising their own cultural and natural values. This task needs to be addressed by institutions at the local and/or regional levels, with an interest in and concern for the bio-cultural diversity of the area.

Historically, the Chamoli highland societies of Garhwal Himalayan have been heritage conscious and it is these people who rose *en masse* against commercial green felling. Many of these community leaders can be regarded as 'folk-ecologists' who possess practical wisdom on *how* to involve local people, *whom* to involve, and *why* to involve indigenous groups in grassroots development programmes. A few of them have their own eco-restoration and eco-development training centres. It is indeed a pity that the resource managers and developers of this region have regarded them as rustic informants and misguided activists.

The remote Bhyundar Valley with its elysian charms of fascinating landscape, floral extravaganza and preserved Himalayan 'little-culture' still retains much of its pristine glory. Tourism in one form or the other has reached there, but much of the beauty has remained unharmed. The planners and developers of tourism have a chance to create a benign form of tourism by practising the concepts and principles of indigenous tourism, discussed in the foregoing. It should be a co-operative development, planned, implemented and monitored by regional and local authorities with a consultative mechanism and proactive participation of informed Bhotias and highlanders. Some sprouting of these seminal ideas has already taken place in Almora of

Kumaun Himalaya where Sauka Bhotias of the Johar region have established a voluntary organization, Jan Jagran Samiti, in Binsar which is struggling for the socio-economic empowerment of women. Similarly, another grass-roots society, ROSE (Rural Organization on Social Elevation) in Kanda (Kumaun) is committed to the integrated development of the rural poor. It also includes rural tourism. ROSE zealously works for the survival of the Kumauni culture and organizes religious events, cultural festivals and traditional arts for the enjoyment of hosts. This movement is likely to take off and encourage more and more Bhotias to practise 'craft-tourism'.

REVISITING THE BHYUNDAR VALLEY

As an epilogue to this theme, it is interesting and perhaps necessary to revisit the valley in the role of a sightseer and to discover the state of the tourism resources and their use and misuse.

Brought into limelight by F.S. Smythe (1931), the Valley of Flowers was known to Hindu sages and saints who found an ideal environment for solitude and meditation. It is also associated with Pandava legend. To the local people, it is a fairy land. In Hindu mythology, it is named Nandan Kanan (the Garden of Lord Indra).

The Valley of Flowers is approachable from Govind Ghat (1,828 m) almost 22 km beyond Joshimath *en route* to Badrinath. About 1.5 km down the motor road one reaches the small settlement, Govind Ghat, offering Sikh hospitality at the *Gurudwara*, and from here the ascent to the Valley of Flowers begins. The trek, about 12.5 km to Ghangaria (the last human settlement) is for the most part along the Lakshman Ganga (mythological name of the Bhyundar). The entire passage is punctuated with fascinating Himalayan sights and sounds, such as waterfalls, roaring rapids, orchards, wild flowers and primitive human settlements with an organic simplicity. The first Bhotia settlement is Phulna, straw-roofed and housed in beautiful surroundings of greens and glades. The next, almost half way, is another Bhotia settlement Bhyundar (2,593 m) named after the river, where one finds Bhotias living in harmony with the natural environment. The ascent beyond this village becomes strenuous, though rewarding, as the scenery dramatically improves. The way begins to reveal the wild beauty of tiny flowers of many colours, of which yellow and pink are the most common. As Ghangaria comes closer, Kharshu and rhododendrons give way to planted stands of majestic deodar.

Bowl-shaped Ghangaria (3,200 m) provides destination facilities of accommodation and food, besides a *Gurudwara* that caters to the needs of Sikh pilgrims to nearby Hemkund, a high-altitude lake, where the Sikh's guru Govind Singh once meditated. The Valley of Flowers is about 3.5 km to the north-east at an average height of 3,650 m, while Hemkund climbs eastward to 4,329 m.

With respect to nature preservation and the biosphere reserve guidelines, the entire Bhyundar catchment can be subdivided into:

1 Cultural Zone: The area beginning from the confluence of the Bhyundar and the Alaknanda up to Bhyundar Village.
2 Buffer Zone: From Bhyundar village to Ghangaria and vicinity.
3 Core Zone: Comprising the entire area from Ghangaria to the Bhyundar Kanta Pass and the Bhyundar Glacier.

The floral aspects of the Valley become perceptible as one leaves Ghangaria. The 3.5 km trek from Ghangaria to the actual Valley of Flowers is idyllic and inspiring. After a strenuous climb of a kilometre, the timberline ends with array of silver birches that herald the green expanse of the Valley. The beauty of the Valley lies exposed before the visitor who has yet to explore the mystery of this 'Garden of Eden'.

The Valley's core spreads into a winding corridor about 8 km long and 2 km across. It slopes up gently to the edge of the Lari Bank Glacier. To the north are impressive cliffs rising over 2,000 m from their base, a rich nesting ground for Himalayan pheasant (monal) and other high-altitude birds. The north-facing slopes are decked with silver birch stands, punctuated with alpine meadows; the south-facing ones are characterized by greenery, snow-covered tops and majestic silhouettes. The most spectacular sight is that of Ratban (6,127 m) to the north-east, which immediately catches the attention of the visitor.

The Valley floor is dense with floral growth, often muddy and rock-strewn, having flowers quite different from the slopes for reason of climate and soil. The first half of the floor ranges from 3,525 m to 3,962 m and is the normal visiting spectrum for tourists. The second half, being sparse and rock-filled, is an area more for field scientists and adventurers and it ends at the base of the Bhyundar Kanta Pass (5,450 m).

The Valley slopes, particularly the south-facing ones, have rare flowers along the runnels, crannies and towards the glacier side, and the Himalayan poppy (blue and yellow) and *Brahm Kamal* (*Saussurea obvallata*) can be found. This is the scene of the Valley's early flowering of primulas and anemones, which gradually works its way to the floor with the coming of rains. The Valley begins to bloom by mid-May and fades out around mid-September. The valley floor grows richer by July and August, when it is not easy to pass through without crushing the tender plants.

Early British botanists identified over 2,500 garden genera, and Joan Margaret Legge, another British botanist, did some collection work for the Kew Botanical Garden of London. A few Indian botanists have further added to the list of plants, but only scattered information is available regarding a complete inventory of plant species and their families.

The synecology of these flowering plants, mostly medicinal herbs, explains how they live in association with each other and how they maintain their symbiosis after interacting with soil and micro-climate, in as much as one dies enlivening the other. The Valley spectrum changes dramatically over the summer weeks and months. The June spectacle is quite different from July or August, and September is an autumnal glory and a swan song.

The trek to Hemkund (Sikh resort) is a corollary to the Valley of Flowers. It takes the hiker to additional sights to those seen along the Valley. Although visitors are not wading through the ocean of flowers as in the Valley, they see the abundance of Himalayan blue poppies and *Brahm Kamal*, with its intoxicating fragrance, is widespread.

The Hemkund lake is not easily reached and is an arduous trek, with the last kilometre demanding more patience and will-power than stamina. Sikh pilgrims brave it with spartan endurance. Beside the lake is a new gigantic star-shaped structure that does not blend well with the natural environment.

The Sikhs opened this area for pilgrimages around 1970. By the 1980s the Valley of Flowers was oversold to tourists, with inadequate infrastructure resulting in environmental damage to the fragile floral resource. In 1982 the Valley was declared a National Park. The public sector focused on the high-spending tourists, and the locals had few opportunities to become involved except serving wayside teas. The better-off Marcchayas and Garhwalis engaged in pony-trekking services and a few took up the role of guides. The porters' market, as usual, was captured by the shrewd Nepali contractors. In due course of time, the more enterprising Sikhs not only modernized facilities in their wayside *Gurudwaras*, but also introduced canned beverages and a fast-food cult, which inhibited the local tea vendors on the main valley trek. A few Bhotias in Bhyundar and Ghangaria (the last settlements) did make some modest efforts to provide accommodation in their homes for trekkers and budget tourists but were not very successful due to a dearth of funds and expertise. The park managers did not care to incorporate the local people in their conservation programmes. Their grazing rights were usurped and they have not been given concessionaire benefits. Much to their chagrin, jobs created by tourism have been given to outsiders. Having lived for many generations with native plants and herbs, the indigenous people have an enviable knowledge of ethno-botany which many researchers have ignored. Thus it is obvious that unlike ACAP experience, the Bhyundar Valley presents a challenge for the development of human resources for harnessing appropriate tourism. It is a task which needs to be addressed in co-operation with regional university tourism training centres in Srinagar (Garhwal), Nainital and Almora (Kumaun) and Garhwal Development Corporations and other sectoral hill-development agencies. The Nehru Mountaineering Institute in Uttar Kashi also has an important role to play in training the Bhotia youth for mountain guides and trek leaders.

The primary focus in such areas should always remain on trek and trail management with effective interpretation of the trek's product – its uniqueness. A beginning should be made with taxonomic explanations of the park's amazing growth cycle of flowers, their intense and restricted distribution, the micro-climatic aspects and other controls which give the floral beauty such a bewildering enchantment. Sadly, most of the park visitors return from the Valley of Flowers disenchanted because of the lack of basic information and park interpretation. Educated unemployed indigenous residents of the region should be recruited for interpretive training, and they would certainly perform effectively and with a sense of pride. The second most important action would be to establish Bhotia lodges – simple, organic, one-room accommodation units, similar in design to the vernacular architecture, having minimum basic facilities of clean water and toilets. The concept of Bhotia lodges for tourists is very close to the French *gîtes* system. A model unit could be placed either in the Bhyundar Village or Ghangaria with local people as operators. Eco-lodges could be set up around the park's core area. The park could serve as an initiative and learning centre for the local people and for their proactive participation. Once this model has demonstrated that it can succeed, the Bhotias of Bhyundar will come forward with their *Ringal* crafts, woollen and other artefacts of local skills as an economic option for improvement in their living standards. This could result in the development of the entire region along similar lines. As observed earlier, much will depend upon the implementation of sustainable tourism development policy, planning framework, monitoring and the response of the local people.

CONCLUSION

Indigenous people, like the Bhotias of Bhyundar, generally live a marginal existence in rather inhospitable regions with few or limited economic opportunities to pursue. Development of organic tourism appears to be a possible option as their habitat generally abounds in landscape aesthetics and cultural exoticism. However, tribal development through tourism poses a threat to the integrity of ecosystems, and especially to bio-cultural diversity and community authenticity. Only community-based tourism with an effective linkage of native resources, such as forestry, horticulture, agriculture and handicrafts, has a chance to survive. Most often other forms of tourism cannot remain sustainable and ultimately decline.

Insular Bhotias with limited community capabilities, coupled with the constraints of poverty, could not take easily to tourism, while their counterparts, the Sherpas of Nepal Himalaya, made the necessity of so doing a virtue by establishing themselves as the world's best mountain guides. Over a brief period they were able to improve their own situation through the practice of mountain tourism, though not without some social costs.

The question that if Sherpas can achieve this, why not the Bhotias, is not easy to answer as they bear striking parallels (clan structures, socio-economic similarities, environmental-deterministic beliefs and political compulsions). One simple reason for the lack of certainty is because no model repeats itself exactly in the Himalaya because of the complexity of its ecosystems and cultural uniqueness.

Nepal has been a more open society than Garhwal or Kumaun, which remained for the most part rather closed systems for strategic reasons. Until the late 1970s even Indian residents had to seek permission to go beyond the 'inner line' of the military. Thus the Bhotia habitat remained inviolate, well preserved and sequestered. By contrast, early interaction of the Sherpas with foreign tourists and mountaineers from across the world moulded their psyche and gave them inspiration for self-development. Germans persuaded them to begin eco-trekking, the Dutch made them environmentally aware while the famed New Zealander Hillary groomed a few of them in park management. These efforts were backed by the political will of their decision-makers who realized that tourism could be a major support for their weak economy. All these factors made Sherpas ambitious and many of them crossed borders to capture markets in Bhutan, India and Pakistan. Mountain research has been conducted by outside funding agencies into the ecological complexity of the Himalaya, and there is now a body of multidisciplinary knowledge on Nepal's high-altitude Himalaya. This has enabled mountain tourism to grow in a scientific manner and the Sherpas to take advantage of this growth. Nepal, in Kathmandu, houses an International Centre For Integrated Mountain Development (ICIMOD) which undertakes research and disseminates knowledge on various aspects of Himalayan ecology and appropriate strategies for human development.

The Indian Himalaya, particularly in its higher reaches, has not been studied seriously, the old British gazetteers provide useful information but they are 200 years old. Some of the places in the greater Himalaya still resemble *terra incognita*. Meagre research information is available on the highland societies: their socio-economic responses, their traditions and mores, their aspirations and their perpetual struggle against a forbidding environment. This is an academic challenge which scholars in the region must accept. There is an onus to provide feedback and information to hill development agencies whose staff remain engaged in administrative programmes with little time for field research.

Tourism has not yet entered Indian Bhot lands, and it should provide an opportunity for a new beginning, supported by well-made plans, based on resource resilience with the consultation of local people. It is very likely that they would favour small-scale tourism which would be allowed to grow gradually and which would be based on a policy of sustainable tourism development.

This new tourism style will be *theirs* – Bhotia tourism, and it will give guests a taste of their landscape, serve them the distinctive food, entertain them with native dances and house them in Bhotia lodges. This new tourism has already began to appear in this part of the Himalaya, and is growing.

Hopefully it will bring some improvement in the standard of living of the Bhotia, while allowing them to maintain the key elements of their culture.

References

Adams, V. (1992) 'Shapers of Nepal: reconstruction and reciprocity', *Annals of Tourism Research* 19 (3): 534–44.

Adu-Febiri, F. (1996) 'Tourism and socio-economic transformation in developing countries', *Review of Human Factor Studies* 11, 1.

Altman, J. (1989) 'Tourism dilemma for Aboriginal Australians', *Annals of Tourism Research* 16 (4): 456–76.

Baumgartner, R. (1992) 'Tourism and socio-economic change: the case of Rolwaling Valley in Nepal', in T.V. Singh, V. Smith, M. Fish, and L. Richter (eds) *Tourism Environment: Nature, Culture, Economy*, New Delhi: Inter India.

Bjonness, I.M. (1983) 'External economic dependency and changing human adjustment to marginal environment in the high Himalaya, Nepal', *Mountain Research and Development* 3: 263–72.

Brandon, K. (1993) 'Basic steps towards encouraging local participation in nature tourism projects', in K. Lindberg and D. Hawkins (eds) *Eco-Tourism: A Guide for Planners and Managers*, N. Bennington, USA: The Eco-Tourism Society.

Butler, R.W. and Hinch, T. (1996) *Tourism and Indigenous Peoples*, London: Thomson.

Chatterji, P.C. (1989) 'Nomadic graziers of Garhwal', in T.V. Singh and J. Kaur (eds) *Studies in Himalayan Ecology*, 2nd edn, New Delhi: Himalayan Books.

Eber, S. (1992) *Beyond The Green Horizon*, Surrey: WWF.

Fantin, M. (1978) *Sherpa Himalaya: Nepal*, New Delhi: The English Book Store.

Gurung, C.P. (1992) 'The Annapurna Conservation Area Project, Nepal', in S. Eber (ed.) *Beyond the Green Horizon*, Surrey: WWF.

Gurung, G., Simmons, D., and Devlin, P. (1996) 'Evolving role of tourist guides: the Nepali experience', in R.W. Butler and T. Hinch (eds) *Tourism and Indigenous Peoples*, London: Thomson Business Press.

Haimendorf, C.V.F. (1979) *Himalayan Traders – Life in Highland Nepal*, London: John Murray.

—— (1981) *Asian Highland Societies in Anthropological Perspectives*, New Delhi: Sterling Publishers.

—— (1984) *The Sherpas Transformed: Social Changes in the Buddhist Society of* Nepal, New Delhi: Sterling Publishers.

Hatcher, R.L. (1996) 'Local indicators of sustainability measuring the human ecosystem', in B. Nath, L. Hans, and D. Devuyst (eds), *Sustainable Development*, Brussels: VUB Press.

Kaur, J. (1985) *Himalayan Pilgrimages and the New Tourism*, New Delhi: Himalayan Books.

Muqbil, I. (1984) 'Trashing of the road to Shangri-La', *Contours* 1: 7.

Niederer, A. (1984) 'Mores, customs and traditions as factors of regional identity', in E.A. Brugger *et al.* (eds) *Transformation of Swiss Mountain Regions*, Bern: Verlag Paul Haupte.

Nityanand, T. and Kumar, K. (1989) *The Holy Himalayas: A Geographical Perspective*, New Delhi: Sterling Publishers.

Parker, B. (1993) 'Aboriginal tourism – from perception to reality', in L.J. Reid (ed.) *Community and Cultural Tourism*, Proceedings of the 1992 Travel and Tourism Research Association – Canada Conference, 4–6 October 1992, Regina, Saskatchewan: Travel and Tourism Research Association, 14–20.

Pawson, I.G. (1984) 'Growth of tourism in Nepal's Everest', *Mountain Research and Development* 4 (3): 237–46.

Price, M. (1996) *People and Tourism in Fragile Environments*, New York: Wiley.

Rodenburg, E.E. (1989) 'Scale in economic development in Bali', in T.V. Singh, H.L. Theuns, and F.M. Go (eds) *Towards Appropriate Tourism: The Case of Developing Countries*, Frankfurt: Peter Lang.

Roughan, J. (1994) 'The Development of Wheel', *SIDT*, Honiara.

Rustomje, N.K. and Ramble, C. (1990) *Himalayan Environment and Culture*, New Delhi: Indus Publishing House.

Sacherer, J. (1981) 'The Sherpas of Rolwaling Valley – a hundred years of economic change in Himalayan ecology', *Ethnologie*, Paris: CNRS.

Singh, T.V. (1989) *The Kullu Valley: Impact of Tourism Developments in the Mountain Areas*, New Delhi: Himalayan Books.

—— (1993) 'Development of tourism in the Himalayan environment: the problem of sustainability', in Ruppert, Rand Storck and Karl-Ludwig *Festschrift Für Wigand Ritter, Zum 60 Geburtstag*, Nürnberg: Friedrich-Alexander Universität.

—— (1996) 'Development of tourism in the Himalaya'. in S. Singh (ed.) *Profiles of Indian Tourism*, New Delhi: Ashish Publishing House.

Singh, T.V. and Kaur, J. (1985) 'In search of holistic tourism in the Himalaya', in T.V. Singh and J. Kaur (eds) *Integrated Mountain Development*, New Delhi: Himalayan Books.

—— (1989) *Studies in Himalayan Ecology*, 2nd edn, New Delhi: Himalayan Books.

Smith, V. (1996) 'Indigenous tourism: the four Hs', in R.W. Butler and T. Hinch (eds), *Tourism and Indigenous Peoples*, London: Thomson.

Sofield, T.H.B. (1993) 'Indigenous tourism development' *Annals of Tourism Research* 20 (4): 729–50.

Sofield, T.H.B. and Birtles, A. (1996) 'Indigenous people's cultural opportunity spectrum for tourism', in R.W. Butler and T. Hinch (eds) *Tourism and Indigenous Peoples*, London: Thomson.

Uppadhaya, V.S. (1988) 'Development of highland societies: a case of central Himalaya', in K.L. Bhowmik (ed.) *High Altitude Anthropology*, New Delhi: Inter India.

Zinder, H. (1969) *The Future of Tourism in the Eastern Caribbean*, Washington DC: Zinder and Associates.

13

TOURISM DEVELOPMENT AND NATIONAL PARKS IN THE DEVELOPING WORLD

Cat Ba Island National Park, Vietnam

Nguyen Thi Son, John J. Pigram and Barbara A. Rugendyke

TOURISM AND DEVELOPING COUNTRIES

At first glance, tourism seems tailor-made for the world's poorer nations, and a growing number of developing countries have placed emphasis on tourism in their development plans. Reasons are not hard to find. A ready market is available for the attractions these destinations can offer; many of them have an appealing climate, combined with exotic scenery and a rich cultural and historical heritage. Land and labour costs are comparatively low and, in the absence of significant mineral production or an export-oriented agricultural sector, tourism is a potential source of foreign exchange and can generate new opportunities for employment and stimulate demand for local products and industries. Tourism is also considered to make improvements possible in the local infrastructure with the provision or upgrading of roads, airports, harbour facilities, accommodation, shopping, entertainment, communications, power and water supplies, health services and sanitation.

However, experience has shown that tourism may not always be the most appropriate form of investment for regions of the developing world. A range of economic, sociocultural, environmental, and political questions have been raised which serve to qualify endorsement of tourism as a strategy for development in the world's poorer nations, among them:

- potential for conflict between tourism and traditional cultures, lifestyles and landscapes, especially when dominated by foreign interests and not integrated into local and regional economies and communities;

- economic costs and benefits of tourism and the extent to which multiplier effects apply (Sherman and Dixson 1991: 97–8; Sinclair and Vokes 1993: 208–9);
- sustainability of predatory forms of tourism without rigorous assessments of environmental impact; and
- antagonistic attitudes of communities affected by tourism in the absence of tangible, compensating socioeconomic benefits (Richter 1993: 194–5).

Even in those developing countries where tourism appears to be fulfilling the promise of economic benefits, foreign exchange earnings are offset typically by considerable expenditure on imports attributable to tourism (Hitchcock *et al.* 1993: 18). Increased purchases to meet the demands of tourists include food and drink, accommodation and outfitting requirements, tourist shopping products, and the necessary infrastructure for international travel (Pattullo 1996: 39). A further drain of foreign exchange can result from repatriation of the profits of offshore corporations and the payment of salaries to higher-skilled personnel brought in to occupy the better-paid positions in the local tourism industry. At the same time, the number of jobs open to the indigenous population may not meet expectations because the local availability of even the limited skills necessary is restricted (Sinclair and Vokes 1993: 206). Many jobs created are temporary or insecure, and may be seasonal, dependent on tourist flows.

Commitment to tourism in developing countries can also bring with it the twin dangers of loss of control over resource allocation and decision making, and dependence on a fluctuating economic base. Tourism is notoriously open to pressures from inflation, fuel crises, industrial troubles, security fears and the fickle nature of resort popularity. If local populations desert traditional occupations for work in tourist undertakings, any downturn in international travel can be doubly unfortunate. Frequently, the rate and direction of tourist development are in the hands of transnational corporations, especially in the areas of transport, hotel chains, tour wholesaling and marketing. This dependency relationship means that much of the profits from tourism in fact leak back to countries in which multinationals are based (France and Towner 1991; Sherman and Dixson 1991: 109; Pattullo 1996: 39). In the South Pacific, for example, no island nation owns any of the companies which operate cruise ships and few governments can afford an independent national airline. The extent of such external control is growing with the tendency worldwide towards vertical integration in the tourist industry (Pearce 1989: 37–8). In the competition between local interests and global capital, local people are generally the losers (Shaw and Williams 1994: 42–3).

Any economic benefits from tourism must also be balanced against possible adverse sociocultural consequences. Costs and benefits are not evenly distributed between local residents and tourists, or within host communities (Whelan 1991: 9; Hitchcock *et al.* 1993: 19; Din 1993: 328). For example, in a recent study of tourism in Nepal, Pagdin (1995) found that tourism only brought improved standards of living to those directly involved in the tourism industry, while associated inflation in fact eroded the living standards of other members of the host communities. Thus, tourism tends to be 'characterized by, and to further engender, social inequalities' (Hitchcock *et al.* 1993: 19).

The confinement of tourists and their expenditures to air-conditioned enclaves offering carefully programmed travel experiences can become a source of frustration and resentment for those local people who do not benefit directly, and may merely accentuate the gulf between affluence and poverty. Imported goods and services compete with local enterprises and conflicts can occur as commercialization of land and resources intrudes on traditional values. Local people may be pressured to leave their lands to accommodate the expanding tourism industry and to allow for the development of new, particularly beach-side, sites (Pleumarom 1992). The nature of the impacts of tourism on the aesthetic and material culture expressed in traditional handicrafts sectors has also been of concern (Parnwell 1993).

Tourism can also serve as a powerful agent of social change and disruption (Shaw and Williams 1994: 14). The excesses of tourism and the unregulated behaviour often associated with it can act as an affront to host cultures (Din 1993: 332; Richter 1993: 198; McCarthy 1994). Conspicuous consumption, eccentric clothing, unacceptable sexual behaviour and illegal activity, including the use of drugs, can all be sources of inter-personal and inter-cultural tensions (for example Pattullo 1996: 84–101). An increase in crime and anti-social behaviour is only one of the many possible undesirable consequences of tourism, even in developed countries, but the extent to which tourism is capable of inducing change in a host society is a function of both the characteristics of the visitors and those of the destination.

Sharp economic and cultural differences between indigenous people and tourists, as is typically the case in newer tourist destinations in the developing world, can foster cultural disruption and transformation of lifestyles. Thus, there have been complaints that tourism results in the 'commodification' of culture; that local cultures are exaggerated and packaged into an 'image' of local culture which is then sold to tourists (Greenwood 1997). Conversely, it has been argued that local cultures and tradition can 'be reconstructed and manipulated and could co-exist with development and modernity', or even that tourism can strengthen local cultures (Wood 1993: 52; 55–6). It has also been suggested that a strong local culture and pronounced nationalistic outlook can act as buffers to distortion of the social fabric (Wilson 1993: 38). While the extent to which social change is negative is debated, and it is obvious that the nature of change is context-specific, there is little doubt that '. . . if cultures travel they cannot be closed-off from other cultures. The admixture of elements and the unintended production of new cultural values are an inevitable consequence of movement' (Rojek and Urry 1997: 11).

In tandem with global interest in the environmental consequences of development, the impacts of uncontrolled development of the tourism industry on the environment have been, increasingly, the focus of research (e.g. Green and Hunter 1993; Parnwell 1993; Buhalis and Fletcher 1995; Pattullo 1996). Commentators argue that the hunger for quick profits, coupled with an astounding lack of planning, have created a 'dark side of paradise' in many destinations in developing nations, with foreign-owned tourism development contributing to marine pollution and destruction of fishing grounds, wetlands, forests, reefs and coastal resources (Pleumarom 1992; Parnwell 1993; Leser 1997: 16). Ironically therefore, tourism development can threaten the very things tourists want to experience – local cultures and environments (Hawkins 1993:185). Thus, 'tourism may be destroying tourism' (Hitchcock *et al.* 1993: 22).

In this context, developing countries could be expected to search for less predatory and more benign forms of tourism to help promote economic development. Ecotourism, or nature-based tourism, increasingly is being put forward as an environmentally and socially acceptable option with potential for economic returns (Whelan 1991: 4). In particular, establishment of national parks is seen as one way of opening up the natural attractions of the Third World to tourists in a sustainable manner. This is also attractive to tourists, many of whom are interested in new and exotic locations and are increasingly searching for the 'otherness' of unspoilt natural environments (Boo 1991: 187–8).

NATIONAL PARKS IN DEVELOPING COUNTRIES

Despite the appeal of national parks as a focus for nature-based tourism, problems can arise with their establishment and management in developing countries. Although ecological considerations and the desirability of preserving unique ecosystems may certainly be recognized in the selection of environments and landscapes for inclusion in the parks system, especially in developing countries, park proposals are often assessed primarily against potential economic and social benefits (Cochrane 1993: 317). This means that, in negotiating land acquisition and planning the future operation and management of a park, it becomes of fundamental importance for the government authority to be able to demonstrate specific benefits, especially for the local people, by way of commercial opportunities and employment (Rovinski 1991: 43). Thus, economic factors may overshadow ecological considerations to the detriment of the park environment (Sherman and Dixson 1991: 90–1).

National parks are now a reality in all corners of the developing world and South-East Asia is no exception. Some of these parks and reserves reflect attempts to protect natural landscapes and wildlife for conservation and scientific purposes (e.g. Olindo 1991; Rovinski 1991; Hitchcock 1993: 310). In other situations, potentially large returns from tourism appear to have influenced their creation. Much of the stimulus for this tourist activity comes from worldwide interest in viewing nature and the diversity of animal and bird life to be found in the national parks. In the less developed countries of Africa, for example, most of the park visitors are from abroad. Whereas some developing nations may regard the parks as unwelcome vestiges of previous foreign dominance, they are tolerated and even encouraged because of their role in providing local employment and attracting tourists and foreign currency (Cochrane 1993).

Although a number of national parks have been established in South-East Asia and the Pacific Islands, considerable difficulties remain to be overcome. Countries like Indonesia, Papua New Guinea, the Philippines, Malaysia and Vietnam apparently have endorsed the concept of national parks and appear convinced of the role they can play in nature conservation. However, such conviction cannot always be assumed in societies where wilderness is still considered an obstacle to progress and the need for conservation is not universally appreciated. There may well be difficulty in diverting money and manpower to the development and management of parks (Olindo 1991: 30) and a reluctance to take land out of what is considered to be more productive use. Even in situations where the authorities do display enthusiasm and an awareness of the value of parks, obstacles may still surface in translating this into adequate levels of funding and appropriate park management (Rovinski 1991: 54–5).

A particular problem can occur in areas of prior human habitation, especially where land is in communal ownership and land-use practices, such as shifting agriculture, timber getting and hunting, may destroy the environment. Of concern in these circumstances is the extent to which new or existing national parks and nature reserves may intrude upon the lives of local residents, leading to disruption of established patterns of land use and of the social fabric (e.g. Pattullo 1996: 119–20). The problem can be met, in part, by raising standards of living above the subsistence levels which contribute to these destructive forms of land use. Moreover, if the local population can receive some tangible benefit from the establishment of a national park, the people may be more prepared to respect and maintain the integrity of the park environment (Whelan 1991: 8). However, as Hitchcock (1993: 314) has demonstrated in one instance, sometimes local residents gain little in economic terms through such tourism development. In practice, as Cochrane (1996: 242) has argued, it is extremely difficult to achieve both the aims of ecotourism at once – to conserve nature, and to improve the welfare of local people. Thus, a fine balance needs to be struck between the creation of a strict nature reserve on the one hand, and a commercially orientated nature-based tourism enterprise on the other. Otherwise, the possibility will remain of resentment and lack of co-operation where a more environmentally compatible, but less rewarding and beneficial type of park system, is imposed on local communities (Drake 1991: 137; Olindo 1991; Whelan 1991: 9). For both the success of the ecotourism enterprise and to ensure that local communities benefit from the tourism development, local people should actively participate in tourism planning (Boo 1991: 189).

A related issue for park establishment and management in developing countries, and one shared with the developed world, is the dilemma of promoting national parks as an engine of tourism, while maintaining the biophysical integrity of the park environment (Woodley 1993: 90). As Cochrane commented:

> The tensions between tourism and conservation have been exacerbated by the failure of the tourism industry and the parks' managers to understand each other. The world of tourism has tended to view the national parks in terms of a tourist attraction, rather than as a means of protecting increasingly rare ecosystems, while the national parks services have traditionally understood the parks in terms of biota and animal behaviour, but not in terms of tourists' needs.
>
> (1996: 238)

In the absence of sound appreciation of park values, emphasis may be misplaced on maximizing visitor numbers in the interests of economic returns, to the detriment of the park itself. Exceeding the limited ecological and aesthetic carrying capacity of parks means that 'ironically, the survival of protected areas may be threatened by the very thing that otherwise protects them—tourism' (Whelan 1991: 11; see also Rovinski 1991: 52).

These concerns were the focus of research into the interaction between tourism and national parks in a recently developed tourist destination in Vietnam. The research was centred on Cat Ba Island National Park in northern Vietnam.

TOURISM IN VIETNAM

Many people's perceptions of Vietnam are coloured by their memories of the war between the south and the north, and involving outside nations. Since the end of hostilities in 1975, reconstruction, bolstered by foreign aid, and the recent liberalization of the economy, are gradually opening up Vietnam to the world. A feature of these changes has been a rapid increase in foreign investment, especially in joint venture operations in tourism projects. In 1993, for example, the hotel and tourism sector received 16 per cent of total investment capital and was ranked third after the industrial sector and oil and gas (Vu Tuan Anh 1994). This, coupled with a gradual relaxation of visa regulations (Lenz 1993), and moves towards a free market economy, have contributed to a growing number of people visiting Vietnam as tourists.

East Asia and the Pacific is the second fastest growing tourist region in the world, accounting for almost 26 per cent of the total increase in international tourist arrivals worldwide between 1995 and 1996 (WTO 1996). Arrivals to the region increased by 7.9 per cent in 1996 to almost 90 million, while receipts totalled over US $82 billion, 13 per cent more than in 1995 (WTO 1996). Despite its central position in this region, Vietnam has not played a significant role in attracting tourists to South-East Asia. In 1992, Vietnam received only 2 per cent of all international tourists arriving in South-East Asia (VNAT 1994). However, this is changing as both foreign visitors and domestic tourists discover the range of natural and cultural features the country has to offer. The availability and standard of facilities for tourists are also improving rapidly. The existence of national parks and many protected areas throughout Vietnam, which act as nature-based tourist attractions, have also contributed to this growth (Figure 13.1).

In a 1993 study it was suggested that, if Vietnam wishes to become a competitor in the world tourist market, and especially in the South-East Asia sector, a number of weaknesses need to be overcome, among them:

- insufficient hotel capacity and poor quality standard of accommodation;
- weak infrastructure and a shortage of foreign capital;
- lack of an educated and well qualified workforce for employment in the tourism industry;
- poor maintenance of historical buildings and heritage sites;
- inadequacy of basic statistical data about tourism;
- the extent of government control and influence;
- over-dependence on foreign tour operators, restricting the scope of the Vietnamese people to choose their own direction for tourism development (de Lozanne *et al.* 1993; Lenz 1993).

Whereas strenuous efforts are being made to address these shortcomings, most of the activity appears to be concentrated in the larger urban centres. It was estimated that, by 1997, there would be ninety-five new or upgraded hotels built to international standard, including ten five-star establishments. However, most of these are expected to be in Saigon and Hanoi, and aimed at the foreign business sector (Tolhurst 1994). Whether isolated tourist destinations, such as Cat Ba Island and its national park, will share in this growth is questionable.

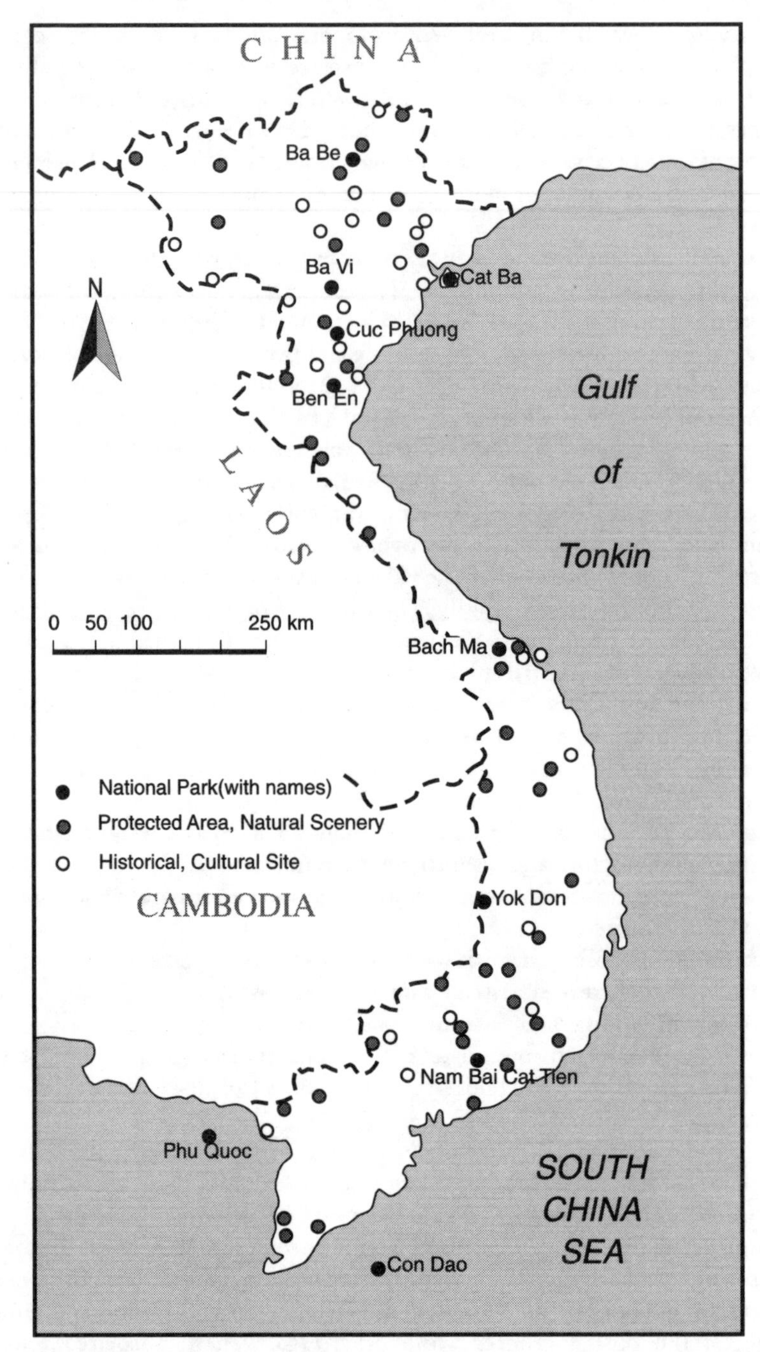

Figure 13.1 Tourism resources of Vietnam.

Source: Map supplied by Cat Ba National Park Office, 1996.

CAT BA ISLAND AND NATIONAL PARK

In the past decade, the number of national parks in Vietnam has grown from two to ten, and many other parts of the country are set aside as protected areas, primarily for scientific research. Cat Ba National Park is considered by many as Vietnam's most beautiful and, increasingly, is a favoured destination with both foreign and domestic tourists. Cat Ba Island is located in the north of the country in Ha Long Bay, 50 km from Hai Phong, the third largest city in Vietnam. It is noted for its karst topography, with many caves and tiny islands, and dramatic landforms (Figure 13.2). A national park was declared on Cat Ba Island in 1986, in order to protect the island's diverse ecosystems.

Although Cat Ba is the largest island in Ha Long Bay, it is only 25 km long and 90 km wide. The national park covers approximately half the island, and includes 90 km^2 of adjacent inshore waters (Figure 2). There is a great diversity of landscapes and ecosystems within the park, encompassing numerous caves, lakes and waterfalls, forests and wildlife. Extensive coral reefs and marine life are found in the coastal waters, along with white sandy beaches and rocky cliffs adjacent to crystal blue seas. Forest cover is also diverse. A unique primitive forest covers 60 per cent of the island, with tropical evergreen forests on lowlands, limestone forests, coastal mangrove forests, and freshwater swamp forests. A variety of fauna exist in the park, including diverse species of mammals, birds, reptiles, and amphibians, and larger marine animals such as seals and dolphins live along small coastal bays around the island. Archeological remains suggest human occupation in some of the cave sites on the island up to 7,000 years ago.

Cat Ba Island enjoys a warm, tropical climate, with the weather divided between a wet hot season and a dry cool season. Heaviest rainfall occurs in the summer months of May to September when tropical storms and typhoons are frequent. From February to April, the weather is often cool and drizzly, but the temperature rarely drops below 10°C.

The many unique and beautiful natural features of Cat Ba National Park offer an impressive range of attractions for those interested in opportunities for nature-based tourism. However, the problems of reaching the island and exploring the national park represent powerful disincentives for visitors, foreign and domestic alike. Moreover, the question remains as to the impact on the relatively undisturbed ecosystems of the park, and whether the island community will welcome and benefit from growing numbers of park visitors.

The people currently living on Cat Ba Island have migrated from nearby coastal provinces on the Vietnamese mainland. The current population of about 12,000 mainly occupies the southern part of the island, including the town of Cat Ba. The population also includes a number of sea gypsies living on their boats. Most of the people make their living from fishing, from

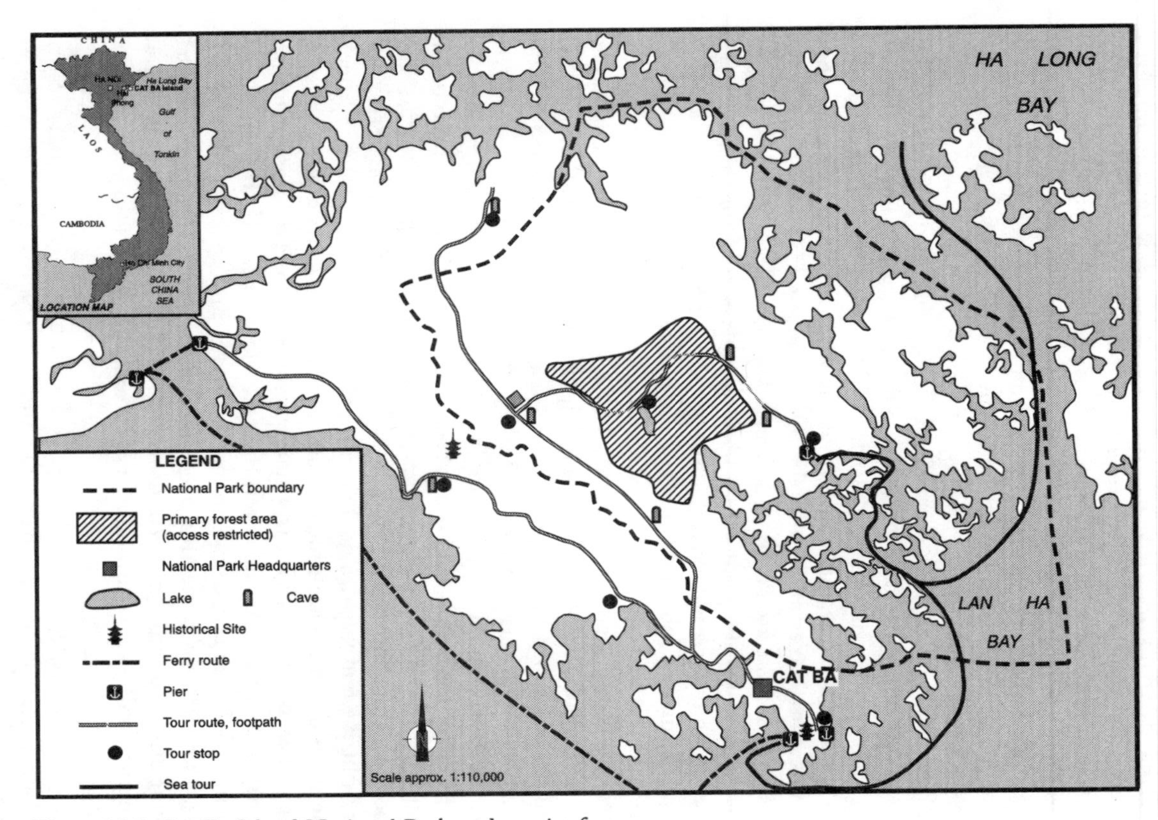

Figure 13.2 Cat Ba Island National Park and tourist features.

Source: Adapted from maps supplied by Cat Ba National Park Office and Cat Ba Tourist Company, 1996.

forestry as a source of pit props for mining operations in nearby provinces, and from agriculture, growing rice, cassava, and fruit. Those living in Cat Ba town operate various types of small, retail businesses including tailors, mini-hotels and motorbike transport.

Some villages and communes are located within the national park, so that agricultural activities and forest clearance take place, and are even encouraged by local authorities. These activities have resulted in deforestation and destruction of natural vegetation and have had a negative effect on water supply and fish production. Some rural planning programmes have been initiated by park management in attempting to overcome these problems.

Cat Ba Island's isolation is made worse by inadequacies in transportation and infrastructure. A journey lasting three to four hours from Hai Phong in an antiquated, uncomfortable daily ferry service across often rough seas is a serious deterrent to many visitors (Figure 13.2). Travelling around the island and to the national park in small groups is by motorbike. There are now some 300 motor bikes on the island. Larger groups of tourists travel by buses belonging to the Cat Ba Tourist Company along the single main road across the island. Routes leading to communes and remote villages are rough and steep, and in a preliminary stage of development.

Infrastructure for tourism is limited, with water supply and communications, for example, poorly developed. Cat Ba Island has not been part of the national electricity grid and electricity until now has been supplied from two generators for four to five hours per day. Connection to the grid is expected to be completed shortly, using a grant from the government and the province, and this should provide electricity continuously to hotels during the tourist season.

Facilities for tourists also have shortcomings. Hotels operated by the Cat Ba Tourist Company are poorly maintained and situated at a considerable distance from the ferry pier. Some are in worse condition than are other hotels, which were constructed recently by local entrepreneurs without official planning. Almost all hotels lack hot water in bathrooms. There is no tourist centre offering advice, supplying souvenirs, or for booking island tours. The one private restaurant suitable for foreign visitors was built only two years ago. Recreation facilities, such as a golf course, or venues for public events or water sports, are unavailable. There is no zoo or similar facility for observing native animals, apart from a small feeding site. Only one rugged, steep nature trail exists through the national park, which attracts mostly foreign adventure visitors, rather than domestic tourists.

Visitors to Cat Ba Island and National Park

The number of visitors to Cat Ba Island and its national park has risen significantly in recent years. In 1991, there were 3,500 arrivals, including

161 foreigners. The number rose to 8,200 in 1993, with 1,900 foreigners, and 15,600 in 1995, of whom 6,100 were of non-Vietnamese origins (Cat Hai Statistical Office 1996). The island and its attractions have become known to foreign tourists mainly through guide books available to visitors to Vietnam.

The majority of foreign tourists visiting Cat Ba Island appear to be independent travellers. Domestic visitors are typically in large groups organized as package tours, while smaller groups are set up for foreigners. The smaller group size has the advantage of reducing impacts and pressure on the infrastructure, local culture and the park environment. It also allows greater flexibility in visits to areas where infrastructure is deficient and offers an opportunity for visitors to stay in local homes.

Whereas tourism on Cat Ba Island has grown substantially in recent years and has brought extensive benefits to the local people and a measure of economic development to the island, it has the potential to bring about undesirable sociocultural change and environmental degradation. Attention now needs to be directed towards identifying the best means of meeting the demands of tourists, while maintaining the integrity of the island's biophysical environment and its people. Participation by the island communities, the managers of the national park, and those with an interest and involvement in tourism, is called for in setting out appropriate strategies for tourism development on the island.

In pursuit of this objective, a number of field surveys were conducted on the island by Nguyen Thi Son in 1996. These focused on the attitudes, motivations, activities and satisfaction levels of visitors to Cat Ba Island and its national park. Local residents and park managers were also surveyed to assess their attitudes to the development and management of Cat Ba Island National Park. Both formal and less structured interviews were used, coupled with participant observation methods. Approximately one hundred foreign and domestic tourists, along with almost one hundred island residents and selected senior park management staff were surveyed.

The results of the visitor survey indicated that the national park was highly valued by respondents, with almost 23 per cent noting it as the most important motivating factor for the visit to Cat Ba Island, and 55 per cent as the second most important factor. Ranking of the attraction of the national park did not differ markedly between foreign and domestic visitors. Visiting Cat Ba National Park itself also ranked highly among activities undertaken by tourists. The results of the survey also suggest a strong preference among foreign tourists for experiencing the local culture, including visits to historic sites and villages.

In general, the satisfaction levels of visitors and their perception of Cat Ba Island as a tourist destination were positive, with natural features and the overall atmosphere highly ranked. Negative perceptions related to pollution, poor transportation and unsatisfactory services, for example, water supply and electricity. Indications of the likelihood of a return visit suggested that Cat Ba has the potential to attract more tourists, especially those seeking to experience the island's natural and cultural features and the national park. This was suggested despite the difficulties posed by the early stage of development of tourism infrastructure and services.

Whereas the results of the survey of tourists suggest that much remains to be done to improve the island tourism experience, this needs to be undertaken sensitively without further degradation of the natural environment or imposition of negative effects on the local community. Cochrane recently suggested that 'one of the most important elements of future tourism development programmes in protected areas is to ensure that the local communities' perceptions of the area in question are ascertained, and that their needs, aspirations and capabilities are discussed' (1996: 258). Thus, the focus of a second survey, the results of which are discussed below, was on the reaction of the island's residents to the presence of tourists and the role they see for Cat Ba National Park.

Cat Ba Island residents

Results of the survey suggest that residents of Cat Ba Island are aware of the attractiveness of the area for tourists and as a place to live. Although a sizeable proportion of respondents (39 per cent) reported having almost no contact with tourists, these were mainly in the farming, fishing, retired and home duties categories, living in communes or villages some distance from the town of Cat Ba and from the route to the national park. The majority of respondents had some contact with tourists in their business, at work, or in accommodating visitors as guests (Table 13.1). These lived in the town where hotels, shopping and tourist activities are concentrated.

Whereas tourism can be an important source of income for some local people in developing countries, as noted earlier, significant economic, social

Table 13.1 Percentage of residents indicating types of contact with tourists to Cat Ba*

Type of contact with tourists	*Sample*	*Percentage***
(a) Having almost no contact	36	39.1
(b) Knowing some as acquaintances	31	33.7
(c) Accommodating guests in my house	15	16.3
(d) Having direct contacts with tourists at work	9	9.8
(e) Earning money from tourists by own business	21	22.8
(f) Other	1	1.1

Source: Field surveys conducted by Nguyen Thi Son.

* Respondents gave multiple responses.

** Percentage of total sample.

and environmental impacts on destination areas and host communities can occur. When the residents of Cat Ba Island were surveyed regarding the impact of tourism on their personal activities, a fairly high percentage of positive effects was reported, although many respondents felt no impact. Overall, the residents considered tourism to be beneficial to them as individuals, with the positive aspects, such as increased income, regarded as outweighing the few slightly negative impacts on their personal activities. A sizeable portion of the sample responded that they did not know about, or were not affected by, tourism. This reflected previous findings that many of the sample had no direct contact with tourism, so had no opinions about the impacts of tourism. Likewise, the activity 'using forest resources', which relates indirectly to Cat Ba Island National Park, was seen as largely unaffected by tourism.

Some interesting differences in reaction to tourism on the island were revealed when survey data were related to gender, age grouping and occupations in the sample. Overall, the impacts of tourism on personal activities were scored more highly by the female group, with shopping opportunities and economic effects seen as having improved. This may reflect the greater responsibility assumed by women in Vietnam in relation to economic issues and family support. Availability of goods and services was also thought to have improved because the greater numbers of tourists has resulted in higher demands for certain types and quality of food. As a result, local people also had access to a wider range of goods and services than would otherwise have been available.

When the survey responses were cross-tabulated with age groups in the sample, most activities were seen as being affected positively to some extent by tourism. Impacts on individuals were rated somewhat differently by the two dominant age groups – 26 to 39 years and 40 to 55 years, 55.4 per cent and 29.4 per cent of the sample respectively. These deviations support the notion that perceptions of tourism change with age, and possibly with length of residence on the island. For example, the older age group suggested that tourism had a negative impact on 'other' activities. It is likely that older people, who have been resident on Cat Ba Island for longer, may be more aware of other negative impacts on their society or culture. Exploration of this issue would provide a useful topic for further detailed research.

In order to examine any differences in attitudes towards tourism from respondents in different occupations, a comparison was made across five broad categories, as follows:

1 teacher and general employees
2 farming and fishing
3 business
4 motorbike drivers and unskilled labourers
5 retired, home duties and others

The results again indicated that island residents perceived an overall positive effect of tourism on their personal activities, but with a clear relationship between respondents' attitudes and occupational circumstances. The most positive attitudes towards tourism were held by residents in groups 3 and 4 who have strong contacts with tourists and earn income from tourism. Residents who were not employed and had no economic interest in tourism – group 5, for example – felt few effects and registered lower scores.

Tourism can also impact more generally on the host community in terms of income and employment, lifestyle, availability of services and infrastructure, and social and environmental concerns. Table 13.2 summarizes the residents' reaction to the impact of tourism on community attributes of Cat Ba Island, with responses ranging from 'much worse' to 'greatly improved'.

The majority of respondents felt that 'jobs/employment' and 'economic prosperity' had improved (80.4 per cent) or greatly improved (10.8 per cent), with slightly greater percentages of respondents indicating that 'economic prosperity' had improved or greatly improved. Roads and transport (60.9 per cent) and 'eating out' (52.2 per cent) were also regarded as being improved. However, over half the respondents considered that 'water and power supply' had become worse as a result of tourism. Other services and social factors

Table 13.2 Impacts of tourism on aspects of community life at Cat Ba Island. (Percentage of respondents) N = 92

Impacts	*Much worse* %	*Worse* %	*No change* %	*Improved* %	*Greatly improved* %	*Don't know* %
Economic factors:						
(a) Jobs/employment	0	0	6.5	80.4	10.8	2.2
(b) Economic prosperity	0	0	0	81.5	15.2	3.3
Social aspects:						
(c) Roads, transport	0	7.6	29.3	60.9	0	2.2
(d) Eating out	0	2.2	36.9	52.2	0	8.7
(e) Water, power	0	52.2	31.5	12.0	1.1	3.2
(f) Price of goods and services	0	14.1	53.3	31.5	0	1.1
(g) Safety/crime	0	1.1	84.8	10.8	0	3.3
(h) Health services	0	0	73.9	17.4	0	8.7
(i) Recreational facilities	0	2.2	47.8	30.4	1.1	18.5
(k) Traditional life/customs	0	0	86.9	8.7	1.1	4.3
Environment factors:						
(l) Vandalism and litter	0	16.3	64.1	0	0	20.6
(m) National Park environment	0	2.2	41.3	9.8	1.1	45.6
(n) Others	0	0	15.2	1.1	0	25.0

Source: Field surveys conducted by Nguyen Thi Son.

were seen as unaffected either positively or negatively by tourism. For some residents, tourism had resulted in negative impacts on the price of goods and services. The environment was perceived as largely unaffected, with a high percentage reporting 'don't know'. This is suggestive of the fact that minimal damage has been caused by the tourism industry, which is still at an early stage of development. It is also likely that there is a low level of environmental awareness or possible indifference among the island community to nature conservation. However, 16.3 per cent of respondents believed 'vandalism and litter' had worsened, while no respondents suggested that there had been an improvement in this area.

Analysis of responses according to demographic characteristics of the sample revealed few marked differences in attitude towards tourism, apart from a stronger negative reaction to 'water, and power supply' and 'vandalism and litter' from females and older residents. As with the perceived impact of tourism on individuals and their activities according to their occupational categories, the strongest reactions came from those occupations closely linked with tourism. Those employed in business or as motorbike drivers had higher positive scores, while those who were not working or not associated with tourism had less positive attitudes to the effects of tourism on community attributes.

From an environmental perspective, there is some suggestion that tourism is regarded as beginning to result in negative impacts. However, the national park environment is not of concern to residents as yet. This might suggest that the concentration of tourism has not reached the point where residents are aware of any problems. It may also lend further support to the contention that local people appear not to be concerned or well informed about the environment.

Despite some negative reactions to tourism, residents surveyed on Cat Ba Island were unanimous in their support for further development. The reasons given related to enhanced economic prospects and job opportunities and the benefits of tourism in improving the social environment and understanding of the local people. At the same time, residents were aware of shortcomings associated with tourism at present and which need to be addressed. These include:

- infrastructure needs such as transport and accommodation, and
- facilities and services (e.g. water, power, information, guide services) to meet the demands of tourists, as well to ensure that local residents are not adversely affected in their ability to access water and power as a result of tourism development on the island.

In addition, a number of residents, presently living outside the centre of tourist activity, expressed an interest in the extension of tourism development to their area.

Summary of resident/tourism interaction

Although the results of the survey of the reaction of island residents to tourism and its effects are not clear-cut, some general observations can be made which may be relevant to other developing countries at a comparable stage of tourism development. It seems clear that local residents agree that Cat Ba Island has potential appeal for tourists and favour encouraging further development of tourism on the island. However, this finding needs to be qualified by the realization that a sizeable proportion of residents have little contact with tourists. For those who do interact with visitors, the overall impact is seen as positive or, at least, that benefits from tourism, especially in economic terms, outweigh the negative effects. At the same time, respondents to the survey recognized that substantial shortcomings in infrastructure, accommodation, transportation and services need to be addressed. Residents also expressed concern that increased tourism was having adverse impacts on their water and power supplies and some pollution and minor environmental degradation were noted by a small number of residents. Of some concern is the apparent lack of awareness of the potential role of the national park in stimulating tourism interest in Cat Ba Island. This lack of awareness, and the possible detrimental effects of increases in tourism on the park environment, were the subject of further consultations with park management on the island.

Management implications for Cat Ba Island National Park

Any plans to promote Cat Ba National Park as a key attraction for tourists need to take into account the attitudes of park management to tourism and their awareness of possible effect of increased numbers of visitors on the park environment. As part of the research project, senior members of the Park Management Board were interviewed to assess their perceptions about tourism on the island and its impact on the national park.

Visiting the island's national park was seen by the managers as the most important attraction for tourists, especially to non-Vietnamese, who, as a group, were noted for their active participation in park activities. When asked about the ways to attract tourists and yet maintain the park environment, the park managers suggested the use of tour guides to reinforce restrictions and regulations as the most effective means of managing visitors. Another approach put forward was the use of brochures outlining restrictions, although this was not considered likely to appeal to domestic tourists. A number of inspection stations have also been set up to monitor park use, and camping and picnicking in the park are not permitted. However, park managers continue to encounter difficulties, as many island residents live within the national park and wish to carry on activities such as farming, logging and hunting.

Whereas tourism brings benefits to Cat Ba Island, it is also seen as contributing to environmental degradation, especially in sensitive areas such as the national park. From a range of possible impacts, park managers identified litter and pollution, vandalism and the condition of the natural environment, as changing for the worse as a result of tourist activities. However, there was uncertainty as to how much change could be attributed to tourism alone, when day-to-day activities of island residents are taken into account.

Although the managers considered that efforts to make tourists more aware of the need for environmental conservation were not keeping pace with increases in the number of visitors, they felt that tourism, as yet, had had little effect on local people and their traditions. On the other hand, attention to transport, accommodation, and tourist services were seen as being in need of improvement. These shortcomings were also identified in the survey of tourists and this agreement provides useful information for policy makers and planners in relation to future development of tourism to the island and the national park.

Whereas management tended to equate the benefits from tourism with revenue from entrance fees and guide services, the managers agreed that moderate and well-planned tourism may help to maintain and even enhance the island's environment and attractions. Moreover, inclusion of the national park in a visit could be expected to strengthen environmental awareness. In this respect, longer-stay visitors were preferred as these were more likely to be identified with nature-based tourism and educational tours of the national park. To encourage this trend, proposals have been made to upgrade accommodation within the park and to establish a facility for viewing fauna, flora and marine life of the island. An environmental education programme is also considered useful for visitors and local residents alike, as a means of assisting to prevent degradation of the island and the national park.

To summarize, the views of park managers are generally in agreement with those of island residents and visitors, in regard to the impacts of tourism. Park management is made more difficult by the number of communes and villages and related agricultural and forestry activities located within the boundaries of the park. Establishment of the park and the associated emphasis on nature conservation could entail disruption of these activities, but local people do not yet have sufficient alternative opportunities for living support. This, coupled with the low educational level of local residents and domestic visitors, and inadequate awareness of the need for environmental conservation, presents a challenge for park managers in supporting stimulation of tourism development while minimizing degradation of the environment. A further issue is the question of government support, since upgrading of the national park and its facilities to cater for increased tourism must compete for funds with efforts to improve the quality of life for island residents generally.

CONCLUSION

The difficulties faced by managers of Cat Ba Island National Park are not unique to the park, the island, or Vietnam. The concerns identified in the surveys of visitors, residents and managers are shared by other communities in developing countries attempting to open up their attractions to tourism, yet seeking to learn from mistakes made elsewhere (e.g. Din 1993; Pagdin 1995). Nature-based tourism appears to offer a way forward which will minimize environmental degradation and disruption to traditional communities, and national parks seem an ideal vehicle for this purpose. However, establishment of national parks and their management for nature conservation can, in turn, curtail pre-existing patterns of land use without bringing the benefits foreseen from tourism. Careful selection and delineation of land for national parks, coupled with sensitive management programmes balancing nature conservation with maintenance of local heritage and cultural values, and closer involvement of indigenous peoples in park management, will contribute to more positive outcomes. Ultimately, the creation of unique forms of nature-based tourism specific to developing countries, and supportive of their economic and social structures, will depend upon concerted efforts to convince host populations to endorse the establishment of national parks as an appropriate means of encouraging growing numbers of visitors to share the special attractions of these regions of the developing world. As others have suggested previously (Wood 1991: 204; Din 1993: 335–6), success in doing this will, in turn, be dependent on the extent to which such developments are 'resident responsive' (Pearce *et al.* 1996: 9) and on the benefits which local people derive from them.

References

Boo, E. (1991) 'Making ecotourism sustainable: recommendations for planning, development and management', pp. 187–99 in Whelan, T. (ed.) *Nature Tourism: Managing for the Environment*, Washington DC: Island Press.

Buhalis, D. and Fletcher, J. (1995) 'Environmental impacts on tourist destinations: an economic analysis', pp. 3–24 in H. Coccossis and P. Nijkamp (eds) *Sustainable Tourism Development*, Aldershot: Avebury.

Cat Hai Statistical Office (1996) *Annual Report 1996* (in Vietnamese), Cat Hai, Hai Phong: Cat Hai Statistical Office.

Cochrane, J. (1993) 'Tourism and conservation in Indonesia and Malaysia', pp. 317–26 in M. Hitchcock, V. King, and M. Parnwell (eds) *Tourism in South-East Asia*, London and New York: Routledge.

—— (1996) 'The sustainability of ecotourism in Indonesia. Fact and fiction', pp. 237–59 in M. Parnwell and R. Bryant (eds.) *Environmental Change in South-East Asia; People, Politics and Sustainable Development*, London and New York: Routledge.

De Lozanne *et al.* (1993) *Study Project Vietnam: Virtues and Vices*, Rotterdam: Elsevier Science.

Din, K. (1993) 'Dialogue with the hosts: an educational strategy towards sustainable tourism', pp. 327–36 in M. Hitchcock, V. King, and M. Parnwell (eds) *Tourism in South-East Asia*, London and New York: Routledge.

Drake, S. (1991) 'Local participation in ecotourism projects', pp. 132–63 in T. Whelan (ed.) *Nature Tourism: Managing for the Environment*, Washington DC: Island Press.

France, L. and Towner, J. (1991) 'Tourism and the Third World' pp. 51–4 in *Conceptual Frameworks in Geography: Case Studies of the Third World*, London: Oliver and Boyd.

Green, H. and Hunter, C. (1993) 'The environmental impact assessment of tourism development', pp. 29–48 in P. Johnson and B. Thomas (eds) *Perspectives on Tourism Policy*, London: Mansell.

Greenwood, D. (1997) 'Culture by the pound', pp. 19–138 in V. Smith (ed.) *Hosts and Guests: The Anthropology of Tourism*, 2nd edn, Philadelphia: University of Pennsylvania Press.

Hawkins, D. (1993) 'Global Assessment of tourism policy: a process model', pp. 175–200 in D. Pearce and R. Butler (eds.) *Tourism Research: Critiques and Challenges*, London and New York: Routledge.

Hitchcock, M. (1993) 'Dragon tourism in Komodo, eastern Indonesia', pp. 303–16 in M. Hitchcock, V. King, and M. Parnwell (eds) *Tourism in South-East Asia*, London and New York: Routledge.

Hitchcock, M., King, V., and Parnwell, M. (eds) *Tourism in South-East Asia*, London and New York: Routledge.

Lenz, R. (1993) 'On resurrecting tourism in Vietnam', *FOCUS* 43 (3): 1–6.

Leser, D. (1997) 'See Bali and cry: the bulldozing of a culture', *Good Weekend, The Sydney Morning Herald Magazine* May 3: 14–25.

McCarthy, J. (1994) *Are Sweet Dreams Made of This? Tourism in Bali and Eastern Indonesia*, Australia: Indonesia Resources and Information Program.

Nguyen Thi Son (1997) 'Nature Based Tourism on Cat Ba Island National Park, Vietnam', unpublished thesis prepared for the degree of Master of Arts (Honours), Department of Geography and Planning, University of New England.

Olindo, P. (1991) 'The old man of nature tourism: Kenya', pp. 3–22 in T. Whelan (ed.) *Nature Tourism: Managing for the Environment*, Washington DC: Island Press..

Pagdin, C. (1995) 'Assessing tourism impacts in the Third World: a Nepal case study', *Progress in Planning*, 44: 185–266.

Parnwell, M. (1993) 'Environmental issues and tourism in Thailand', pp. 286–302 in M. Hitchcock, V. King, and M. Parnwell (eds) *Tourism in South-East Asia*, London and New York: Routledge.

Pattullo, P. (1996) *Last Resorts: The Cost of Tourism in the Caribbean*, London: Cassell.

Pearce, D. (1989) *Tourist Development*, 2nd edn, New York: Longman.

Pearce, P., Moscardo, G., and Ross, G. (1996) *Tourism Community Relationships*, Oxford: Pergamon.

Pleumarom, A. (1992) 'Course and effect: golf tourism in Thailand', *The Ecologist* 22 (3): 104–10.

Richter, L. (1993) 'Tourism policy-making in South-East Asia', pp. 179–99 in M. Hitchcock, V. King, and M. Parnwell (eds) *Tourism in South-East Asia*, London and New York: Routledge.

Rojek, C. and Urry, J. (eds) (1997) *Touring Cultures: Transformations of Travel and Theory*, London and New York: Routledge.

Rovinski, Y. (1991) 'Private reserves, parks and ecotourism in Costa Rica', pp. 39–57 in T. Whelan (ed.) *Nature Tourism: Managing for the Environment*, Washington DC: Island Press.

Shaw, G. and Williams, A. (1994) *Critical Issues in Tourism: A Geographical Perspective*, Oxford and Cambridge: Blackwell.

Sherman, P. and Dixson, J. (1991) 'The economics of nature tourism: determining if it pays', pp. 89–131 in T. Whelan (ed.) *Nature Tourism: Managing for the Environment*, Washington DC: Island Press.

Sinclair, M. and Vokes, R. (1993) 'The economics of tourism in Asia and the Pacific', pp. 200–13 in M. Hitchcock, V. King and M. Parnwell (eds) *Tourism in South-East Asia*, London and New York: Routledge.

Tolhurst, C. (1994) 'Large hotel groups move to get a share of boom in tourism', *Australian Financial Review* 17 May: 44.

Vietnam Administration of Tourism (VNAT) (1994) *Annual Report* (in Vietnamese), Hanoi.

Vu Tuan Anh (1994) *Development in Vietnam: Policy Reforms and Economic Growth*, The Institute of Southeast Asian Studies, Singapore.

Whelan, T. (1991) 'Ecotourism and its role in sustainable development', pp. 3–22 in T. Whelan (ed.) *Nature Tourism: Managing for the Environment*, Washington DC: Island Press.

Wilson, D. (1993) 'Time and tides in the anthropology of tourism', pp. 32–47 in M. Hitchcock, V. King, and M. Parnwell (eds) *Tourism in South-East Asia*, London and New York: Routledge.

Wood, M. (1991) 'Global solutions: an ecotourism society', pp. 200–6 in T. Whelan (ed.) *Nature Tourism: Managing for the Environment*, Washington DC: Island Press.

Wood, R. (1993) 'Tourism, culture and the sociology of development', pp. 48–70 in M. Hitchcock, V. King and M. Parnwell (eds) *Tourism in South-East Asia*, London and New York: Routledge.

Woodley, S. (1993) 'Tourism and sustainable development in parks and protected areas', pp. 83–96 in J. Nelson, T. Butler and G. Wall, *Tourism and Sustainable Development: Monitoring, Planning and Management*, University of Waterloo: Heritage Resources Centre.

World Tourism Organisation (1996) *Tourism Highlights, 1996*, Madrid: World Tourism Organisation.

14

ENVIRONMENTAL IMPACT ASSESSMENT FOR TOURISM

A discussion and an Indonesian Example

Patricia Simpson and Geoffrey Wall

INTRODUCTION

In recent decades there has been a proliferation of studies of the impacts of tourism. Many of the findings of these studies have been contradictory and the case study approaches which have often been adopted have yet to lead to the cumulative knowledge or level of generalization desired by decision makers (Wall 1996). At the same time, growing concern as to negative effects of developments of all kinds have caused the institution of legal requirements for the completion of environmental impact assessments (EIAs) for major projects, including those for tourism, as a part of the process for gaining project approval. However, with a limited number of exceptions (e.g. Pearce 1989; Butler 1993; Hunter and Green 1995), there is little guidance available in the literature concerning the conduct of formal EIAs for tourism. Furthermore, EIAs require the assessment of possible impacts prior to their occurrence whereas academics have tended to examine impacts of tourism once they have occurred. Thus, there is a disjunction between academic research, the needs of consultants charged with the task of preparing formal EIAs and the requirements of the decision makers whom they are to inform. The situation is further complicated by the fact that tourism possesses characteristics which compound the difficulties associated with the assessment of impacts.

The purposes of this chapter are to draw attention to the need for the establishment of guidelines for the conduct of EIAs for tourism and to illustrate this need through a discussion of the situation with respect to EIA in Indonesia. While it is acknowledged that the forms which tourism development will take will result in different consequences, and the contexts in which they occur may give rise to different assessments of the importance of specific impacts, the problems associated with EIAs are sufficiently generic to be of concern in both developed and developing countries. However, the political, legal and economic as well as the natural environments in which tourism occurs have consequences for the implementation of EIA.

ENVIRONMENTAL IMPACT ASSESSMENT

EIA has been defined by the Canadian Environmental Assessment Research Council (1988, quoted in Doberstein 1992: 12) as:

> A process which attempts to identify and predict impacts of legislative proposals, policies, programs, projects and operational procedures on the biogeophysical environment and on human health and well-being. It also interprets and communicates information about those impacts and investigates and proposes means for their management.

The inclusion of impacts on human well-being indicates that social impact assessment (SIA) is fundamentally part of the EIA process. The definition also indicates that EIA can be useful both in analysing specific projects and as a tool at the planning and policy levels of development, and that it may provide a framework for management of impacts. At the project level, Werner (1992) suggests that EIA can be used as a decision-making tool in determining the acceptability of a project, or as a planning tool to minimize negative impacts of an already accepted project.

EIA processes vary greatly from one place to another, but at the project level, most follow four basic principles (Roberts and Hunter 1992):

1 They identify the nature of the proposed and induced activities which are likely to be generated by a project or the introduction of a process.
2 They identify the elements of the environment which will be significantly affected.
3 They evaluate the initial and subsequent impacts.
4 They are concerned with the management of the beneficial and adverse impacts that are generated.

Alongside current, widespread acceptance that environmental planning is an essential part of development, and that institutionalized EIA is necessary, is the recognition that EIA and its methods are flawed. Biswas (1992a) and Hunter (1995a) discuss many of the problems with formal EIA procedures. There is a tendency for EIA to focus on physical impacts and neglect social and cultural ones, often resulting in the production of overly mechanistic reports that deal almost exclusively with the presentation of data, rather than its analysis. For example, waste or emission concentration levels may be provided with an emphasis on whether acceptable limits have been exceeded, rather than an evaluation of their likely consequences for human or ecosystem health. EIA often focuses upon mitigating negative impacts, rather than attempting to increase beneficial impacts, and compliance monitoring is seldom performed. Assessments often delay developments and cost more than expected, sometimes because they are not undertaken in a timely manner and are not well integrated into the project cycle. And, because it is often narrowly focused, EIA often looks only at the direct impacts of a new development, and not its addition to the cumulative impacts of development in the area.

EIA in developing nations

One of the flaws in EIA methodology is that it is difficult, if not impossible, to transfer procedures from developed to developing countries (Doberstein 1992). Many developing nations now recognize the need to avoid large-scale, negative environmental impacts of development and have begun to institute formal EIA requirements, in part as a response to pressure from external forces such as development aid agencies, foreign governments, and international environmental organizations. The World Bank now refuses loans for development projects that do not include comprehensive EIA documentation (Mieczkowski 1995).

As EIA is becoming widely utilized internationally as a development control procedure, guidelines specific to lesser developed nations are required, and the literature contains several studies and recommendations. Generally, the least expensive EIA procedures, requiring a minimum of technology and resources, are most appropriate in developing nations (Doberstein 1992). Checklists and matrices are examples of methods that allow identification of impacts without incurring great expense or requiring extensive technology or technical expertise for their application (Hunter, 1995a).

Doberstein (1992) summarizes some of the more common constraints to EIA in developing countries. Many of the difficulties he lists are likely a result of lack of experience, due at least in part to the relatively short time that most of these nations have been incorporating aspects of environmental planning into development. Many developing countries do not yet have well-developed institutions to support environmental management, or a history of the application of environmental law. There is often only a small number of skilled workers with the expertise to perform EIA , and little financial incentive for them to work in the public sector. Many poorer countries have not yet collected baseline environmental data, and their bureaucrats may not have an appreciation of its value for EIA. Other problems mentioned by Doberstein include a lack of public awareness of environmental problems, insufficient SIA techniques, and the fact that projects are often planned and undertaken by individuals from one dominating discipline, leading to EIA reports that do not represent inter- or multidisciplinary concerns.

Perhaps one of the most important perspectives to consider in EIA for developing nations is to acknowledge that development must and will occur. Biswas (1992a: vii) summarizes the view from the South, that '. . . environment must not be ignored but development must not be impeded'. While such an attitude may imply that almost every development proposal is accepted (Biswas 1992b), EIA can still be very useful in influencing the quality of development and in mitigating negative impacts. Mitigating measures may be suggested that are more economical to incorporate in the design stage rather than in later stages of the project cycle (Hunter 1995a) or alternative, more suitable, locations may be proposed. Clearly, though, EIA procedures and environmental legislation that do not allow unsound developments to proceed unchecked are desirable.

Social impact assessment should be an extremely important component of EIA, especially in developing nations, because local populations frequently have little voice in decision making. Power is often centralized, with little or no input requested from individuals affected by development (Doberstein 1992). As well, due to high population densities in the southern countries, almost any development will involve some displacement of persons (Dick and Bailey 1994). SIA should ensure that such impacts on humans are considered and managed, and so is potentially a powerful tool for more equitable decision making in developing countries. Most writers agree that environmental education for the local populations is one of the most important components that should be included as part of the SIA process (Doberstein 1992). However, the EIA literature does not suggest a specific format for SIA in developing nations, but advises that any format will need to be adapted to reflect the characteristics of individual societies.

ASSESSING THE IMPACTS OF TOURISM

EIA should be based upon a thorough understanding of the nature of the agent of change if the full implications are to be foreseen. Unfortunately, tourism has a number of characteristics which make it particularly challenging for the conduct of EIAs. Tourism is an extremely complex phenomenon. For example, the tourism industry is fragmented, involving both multinational corporations and a multiplicity of small and intermediate-sized operations interacting in a web of institutional interrelationships: it is an example, *par excellence*, of the intricate links between interacting phenomena operating simultaneously at both global and local scales. These linkages involve operators in both the private and public sectors, and span a diversity of economic phenomena, such as transportation, hotels and restaurants, attractions and shopping purchases, which are not always considered as being part of the same economic sector and whose roles in tourism may be difficult to separate from their other functions (Smith 1988).

It is also somewhat artificial to separate the consequences of tourism into environmental, economic and social domains as is frequently done. This division is usually adopted for convenience but, in reality, they are not separate domains. Recognition of such issues, which are not restricted to tourism, has resulted in the incorporation of SIA's and economic evaluations as components of many EIAs.

EIAs are usually conducted for specific developments such as new resorts. As such, the focus of the EIA may be restricted to the confines of that resort development. However, most tourists do not remain within the resort. They arrive by air or other form of transportation and must be transported to the resort. Thus, the new resort has implications for the number of jets arriving at the airport and the number of taxis and buses on the road. They also travel to see the sights in the vicinity and thus penetrate other parts of the destination region. Such situations are both difficult to document and result in challenges in drawing up the terms of reference for the conduct of a tourism EIA to ensure that it will encompass both on-site and regional impacts.

While large developments are natural candidates for EIAs, the cumulative impacts of many small developments may be just as troublesome but much more difficult to encompass within traditional EIA processes. Tourism exhibits many of the characteristics of common property resources, where there may be an incentive for individual entrepreneurs to expand their operations to the detriment of others, resulting in the degradation of the resources on which they all ultimately depend (Hardin 1968). The gradual, insidious development of a multitude of small accommodation units, restaurants and souvenir outlets can rapidly change the character of a place but it is time-consuming and expensive to conduct assessments of every minor initiative.

It is easy to write of tourism as if it were an undifferentiated phenomenon. However, tourism varies greatly in scale, environmental setting and activities undertaken. The consequences of tourism will also be modified by the policy context and the roles which intermediaries, such as tour guides, play in influencing interactions between visitors and local people. This makes it difficult to adopt class assessments as is sometimes done for some sectors and activities, such as forestry or road construction, where the lessons learned from one development may be readily applied to another.

It is desirable that both the negative and positive consequences of tourism be assessed prior to development in order that undesirable effects can be avoided or mitigated, and the desirable effects enhanced. Unfortunately, the literature on impact mitigation as it might be applied to tourism is extremely limited (although see Long 1992). To complicate matters further, tourism is often directed at special environments where the mitigation of adverse environmental changes may be particularly difficult to address. High energy environments, such as coasts and mountains, are often sought by tourists. Tourists are often not satisfied with experiencing usual situations but wish to see special buildings, cultural festivals or endangered species, making their potential for disruption particularly marked.

The impacts of tourism are well-documented and it is widely acknowledged that tourism can potentially affect virtually all aspects of the environment, and that the social and cultural impacts of tourism may be profound. Mathieson and Wall (1982), Mieczkowski (1995), and Shera and Matsuoka (1992) discuss the social and cultural impacts of resort developments on local communities. The physical changes to the area will affect the way of life of local populations, while interaction with tourists may have an enormous influence on the entire society. No other kind of development includes the anticipation of a continued influx of outsiders, who are not expected to try to integrate with the local community, and who will interact with them in such a wide variety of situations. As well, the resources devoted to tourism developments are no longer available for the traditional users, which may cause inconvenience or even hardship.

Yet in many places tourism has been allowed to develop without being previously evaluated by EIA processes. Furthermore, while SIA is incorporated into most current EIA processes, it is a traditionally an area of weakness (Hunter 1995a). Although current tourism literature supports the use of EIA in evaluating tourism developments (Ceballos-Lascuráin 1996), EIA literature does not make much mention of tourism, nor does it address the unique nature of its impacts.

In 1982, Mathieson and Wall (p. 5) wrote that, due to the newness of EIA, there was a paucity of methodological guidelines for undertaking investigations of the impacts of tourism. Now, fifteen years later, though guidelines for EIA methods are plentiful, few are specifically targeted to assess the impacts of tourism developments. Mieczkowski (1995) suggests that such studies should be made comparable with one another through the use of similar methodologies, scales, and levels of analysis. Hunter (1995a) suggests some general criteria for use in determining the necessity of EIA, recommends that EIA should be performed for all planned and pre-existing tourism developments, and postulates that they will be most successful if they are performed within the context of a national framework for balancing development goals and environmental concerns.

THE INDONESIAN RESEARCH CONTEXT

Many developing nations lack a comprehensive framework for environmental planning and management, so little is done to prevent, or mitigate, the resulting environmental degradation. This is not the case in Indonesia where

EIA legislation has existed for approximately a decade. We now turn to a discussion of the Indonesian situation and, specifically, to the application of EIA to a resort development in North Sulawesi. The work which is reported is based upon a detailed review of the EIA literature in general as well as that for Indonesia, assessment of the legislation and regulations for EIA in Indonesia, familiarity with tourism in Indonesia gained from almost a decade of research on varied aspects of tourism in that country, and detailed assessment of the application of EIA for several tourism developments in North Sulawesi. A brief overview of tourism and EIA in Indonesia will be presented prior to the discussion of a detailed case study of the Paradise Beach Hotel and Resort in Likupang, North Sulawesi. In this way both theory (what should be done) and practice (what is actually done) will be addressed.

Tourism and development in Indonesia

Indonesia is an example of a developing country with a burgeoning tourism industry. Until the 1996–7 financial crisis, the nation's productivity and economy had improved greatly over the preceding decades, mostly due to the exploitation of natural resources, but the environment displays evidence of the costs of these advancements (see, for example, for Bali, Martopo and Mitchell 1995 and Knight, Mitchell and Wall 1997). Among other effects, air and water pollution are severe, especially in urban centres; massive cutting and burning is decimating the tropical rain forest; and rapid change and increasing urbanization are affecting the people and their way of life (World Bank 1994). The government of Indonesia is aware of the need for environmental management within development plans and has passed environmental regulations, thus making the country an interesting case study for examination of the theory and practice of EIA for tourism developments.

The national development strategy of Indonesia is to promote growth, equity and sustainability – a trilogy viewed not as three separate goals, but as 'mutually interdependent aspects of the same ultimate goal: improving the quality of life for all of Indonesia's citizens' (World Bank 1994: 7). One of the Indonesian government's strategies for increasing equity within development is to target areas of high poverty, both in the more developed western provinces and in the more remote communities of the less developed eastern provinces. Tourism is a principal tool in the attempt to expand economic growth in Eastern Indonesia (Wall 1997).

Much of the tourism in Indonesia is based on coastal resources, and the national government's current five-year development plan (*Repelita VI*) supports expansion of marine-related enterprises including tourism, especially in the eastern islands of the archipelago. One of the specific targets of the development plan is to have an annual growth rate of 12.9 per cent in tourism, and it is acknowledged that most of the tourists will be attracted to the coastal environments (Ministry of State for Environment 1996). Prior to the recent economic crisis, tourism was growing just below the target rate at about 11 per cent per annum, had become a leading source of foreign exchange in the country, and was also in the top three industries in terms of contribution to gross national product (Telfer 1996).

Indonesia's National Tourism Strategy concentrates on beach tourism, with high quality accommodation aimed at high-expenditure visitors as a development priority, although there is also recognition of the need for medium-level facilities for domestic tourists (Richter 1993). The parts of the country visited most by international tourists are presently the islands of Java and Bali, and a major goal of the strategy is to sell the concept of add-on visits to other parts of the country (Directorate General of Tourism/UNDP 1992). The city of Manado in North Sulawesi is one of several centres in Eastern Indonesia targeted for luxury developments (Wall 1997).

At the same time, the national government is committed to 'issues of global sustainability' (The World Bank 1994). Indonesia's 1982 national environmental laws commit to the principles of environmental sustainability, and subsequent legislation has reinforced that commitment. As the tourism industry in Indonesia is encouraged to grow and has become one of the driving forces of regional economic development, it is necessary to ensure that it not only contributes to regional and national incomes, but also that it improves the quality of life for the people it affects. The National Tourism Strategy recommends that tourism developments must conform to high standards of environmental care, and that they must consider the social and cultural characteristics of the host communities (Directorate General of Tourism/UNDP 1992). With the rapid expansion of the industry, it is becoming increasingly important to create environmentally sound tourism that contributes to sustainable development.

EIA in Indonesia

The history of environmental legislation in Indonesia has been documented by Dick and Bailey (1994) and Conover and Hanson (1992). While EIA has been performed informally in Indonesia since the early 1970s, the first guidance for implementing formal EIA procedures was regulated in 1986 and updated in 1993 with the currently valid Government Regulation Number 51 of 1993, 'Regarding Environmental Impact Assessment'. In this national regulation, environmental impact assessment, or AMDAL (*Analisis Mengenai Dampak Lingkungan*) as it is referred to in Indonesia, is defined as 'the process of studying the significant impact of a proposed business or activity on the environment, which is required as part of the decision-making process' (BAPEDAL 1996: 48). The regulation requires AMDALs to be undertaken for planned developments that are expected to have significant impacts on the environment, while significant impacts are defined as fundamental changes to the original environmental profile of the area in question. Fundamental changes are defined as those with 'either positive or negative impacts on the environment so as to either facilitate or complicate the achievement of the objectives of environmental management' (BAPEDAL 1996: 74).

In 1987 several generic guidelines for AMDAL were established, and the Minister of the Environment, along with the government agency responsible for managing environmental impacts (BAPEDAL), has since continued to develop AMDAL procedures. In 1994, the government department responsible for the tourism sector, Depparpostel (Ministry of Tourism, Post and Telecommunications), released guidelines specifically aimed at assessing the impacts of tourism developments on the environment.

Since the 1986 legislation, the AMDAL process has been implemented by various government departments, as well as for private developments. The legislation is comprehensive in that it makes the application of EIA mandatory for all new developments in both the private and public sectors which are expected to have significant impacts. Several central government agencies and the provincial governments share responsibility for administration of AMDAL, depending on the level and sector of government responsible for individual development projects.

The AMDAL process is outlined in national regulations. Regulation 51/93 emphasizes that integration of environmental management into national policies on growth and development is vital. It stipulates that AMDAL should take place as part of the feasibility study of a proposed project, it makes provisions for public participation, and it determines that projects must develop environmental management and monitoring plans as part of the AMDAL process. It includes assessments of social impacts as an integral part of the process.

Briefly, all AMDALs consist of four main stages:

- scoping, documented in the Terms of Reference (KA-ANDAL)
- impact analysis, documented in the Environmental Impact Statement (ANDAL);
- development of an Environmental Management Plan (RKL), and
- development of an Environmental Monitoring Plan (RPL).

The process begins with scoping, to determine both the aspects of the project that are likely to have significant impacts, and the environmental components that are likely to be affected. The KA-ANDAL document contains the results of the scoping process, and serves 'as a reference to improve the efficiency and effectiveness of the environmental impact statement preparation process' (BAPEDAL 1996: 70). As well, the project, ecological, social and administrative boundaries within the AMDAL study area are stated in the KA-ANDAL.

The AMDAL process then continues with the ANDAL, 'a detailed and in-depth research study on the significant impacts of a proposed business or activity' (BAPEDAL 1996: 49). After the analysis is complete, the RKL and RPL are developed. The RKL presents 'those efforts that will be made to manage the significant environmental impacts which will result from a proposed business or activity' (BAPEDAL 1996: 49), and must both outline how the impacts outlined in the ANDAL will be managed, and clearly identify the party or parties responsible for performing the environmental management. The RPL presents 'those efforts that will be made to monitor the environmental components which will be subjected to significant impacts arising from a proposed business or activity' (BAPEDAL 1996: 49). It formulates a long-term plan, including methods of impact monitoring and, as in the RKL, should indicate the party or parties responsible for monitoring activities.

To avoid confusion, it should be noted that the term AMDAL is used to cover the entire process from scoping to monitoring, whereas ANDAL refers to a specific part of the process, the Environmental Impact Statement.

Regulation 51/93 states that the AMDAL process must be performed as part of the feasibility study for project proposals, for use in the decision-making process. If the ANDAL determines that the negative impacts of the proposed project are impossible to mitigate, or that the cost of managing them will outweigh the positive impacts, then the project shall be rejected. Further, no final operating permits are to be granted until the RKL and RPL have been not only approved, but also implemented (BAPEDAL 1996).

In reality, the AMDAL process is rarely executed in the manner outlined in the country's well-developed EIA legislation and regulations. Interviews with AMDAL consultants and government officials reveal that a variety of hindrances, from corruption, to inefficiency, to a lack of environmental expertise, to uncertainty about responsibilities, contribute to the frequent production of hurried or superficial AMDALs. Some of these issues will be illustrated in the case study which follows.

PARADISE BEACH HOTEL AND RESORT, NORTH SULAWESI

Sulawesi is one of Indonesia's eastern islands with a rapidly expanding tourism market. It has both cultural and natural attractions, including scuba diving at the internationally renowned Bunaken National Park, as well as a number of rare endemic animal species which are protected in parks and wildlife reserves (Whitton, Mustafa and Henderson 1987). Both North and South Sulawesi are in the 'development stage' in Butler's model of tourism destination cycles of evolution i.e. a stage of rapid growth (Sofield 1995). Kinnaird and O'Brien (1996) report that international tourism in the province increased by 400 per cent between 1990 and 1996, and that visitation has been growing by 30 per cent per year.

The North Sulawesi provincial plan for regional tourism development supports the national plan for development in Eastern Indonesia, and large parts of the coastline are slated for further tourism development, including the most remote tips of the peninsula where few other industries operate (Pt Dinnelator Likupang Beach Paradise 1997). Such places are the sites of numerous coastal villages where the people often have a subsistence-level existence, based largely on marine resources and agriculture. In Sulawesi, as in much of Indonesia, coastal communities are characterized by low levels of formal education, limited capital and technology, little access to economic infrastructure, few social services, seasonal unemployment when monsoons prevent fishing, and over-reliance on fishery resources (Ministry of State for Environment 1996). Alternative and supplementary employment opportunities are badly needed but, as Carter (1997: 48) points out, tourism development in Indonesia tends to conflict with traditional uses of coastal resources and the benefits of development 'do not always trickle down to coastal communities that sacrifice access to coastal resources in the name of development'.

North Sulawesi, with its rapidly growing tourism industry, has several recent developments for which AMDALs have been performed, and is a logical location for research into the use of EIA. The Paradise Beach Hotel and Resort is a development which has undergone the AMDAL process. The AMDAL documentation was obtained, field observations were made and interviews were conducted with individuals involved in conducting the AMDAL, local government officials, as well as those affected by or familiar with the resort. Thirty interviews were also completed with residents of Maen and Tamba, villages in close proximity to the resort. Information was collected on how the AMDAL process had been applied to the project, about the environmental impacts that had occurred as a result of the project, and about how these impacts are being managed and monitored. Investigations were undertaken between July and September 1997.

Population, economy, and setting

The Paradise Beach Hotel and Resort is in the district of Likupang, a remote area on the tip of the North Sulawesi peninsula, where tourism development is a major priority (Figure 14.1). The hotel opened for business in 1995 and employs approximately 250 people, while construction of more rooms, staff housing, and a golf course, is ongoing. Paradise currently consists of a luxury hotel with swimming pools, tennis courts, a golf course nearing completion, a jetty with a diving operation, and a small nature reserve with a jogging trail. However, with a project site of 450 ha, the long-term plan is for a much larger resort. By the year 2006, the site should have at least two more

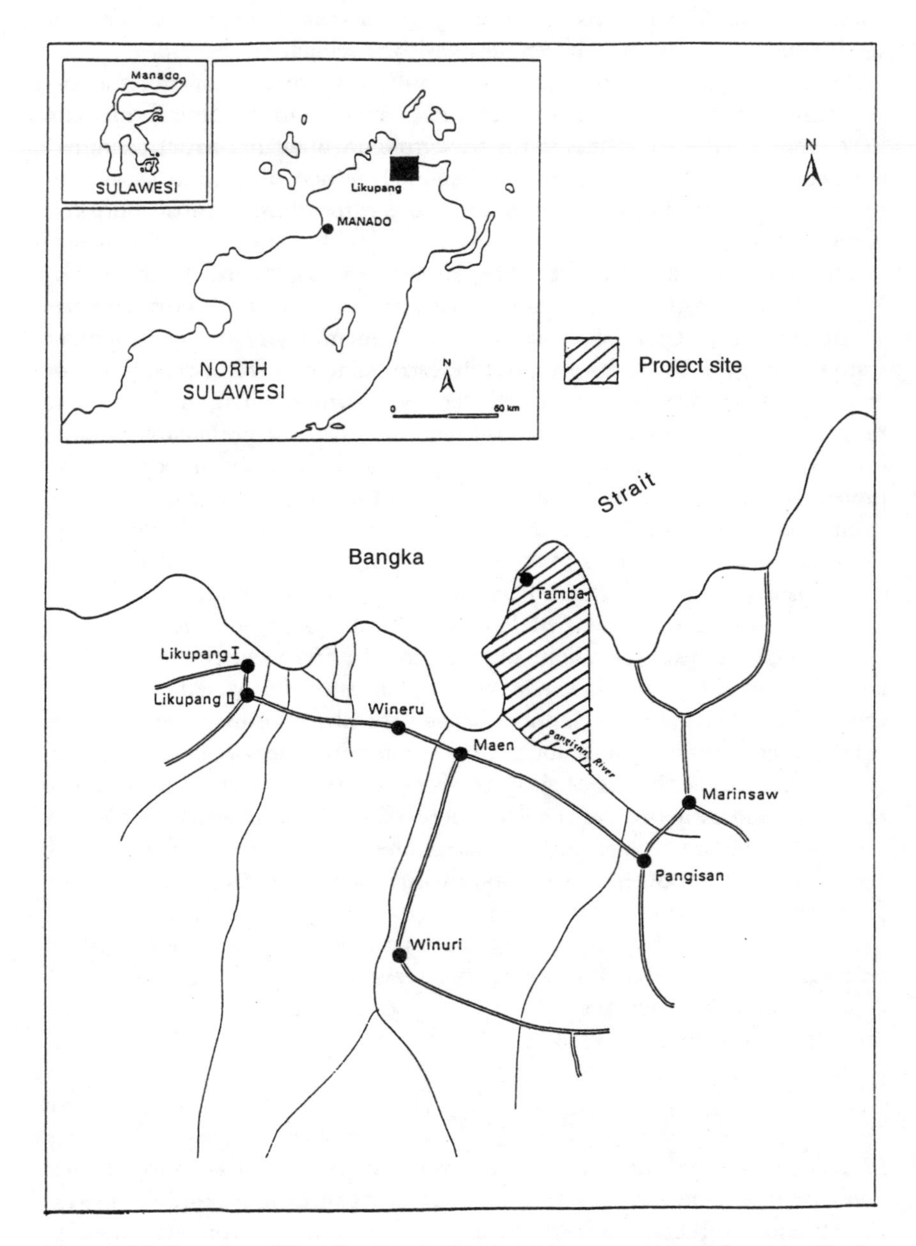

Figure 14.1 Location of Paradise Beach Hotel and Resort project, Likupang, North Sulawesi, Indonesia.

hotels, retirement apartments, campgrounds, more golf courses, a marina, a floating restaurant, a bird park, a fruit orchard, a shopping centre, and even a small village with its own church, hospital, and school.

The main attraction for tourists is scuba diving. The site is very attractive naturally, with both white and black sand beaches, dramatic rock formations in the water, and rolling hills in the background which are currently covered with coconut palms. The diving and snorkeling operation is in an area of healthy coral reefs and off-shore islands, and extends into nearby Bunaken National Park.

The population density in the area is quite low by Indonesian standards. Maen, the village closest to Paradise, has just over 900 inhabitants and the administrative district (*Kecamatan*) has a population of 32,400 in several scattered villages. There is very little industry aside from some production of copra, and over 75 per cent of the residents are farmers, with fishing the next most common occupation. Most villagers catch fish for personal consumption and also grow their own produce. Approximately half of the land at the project site is fertile and suitable for agriculture, including corn, coconut palms, household gardens and raising of domestic animals, while the rest of the land is swampy or has poor soil that can support only scrub and grasses (Pt Dinnelator Likupang Beach Paradise 1997).

Prior to the development of Paradise, Tamba, a sub-district of Maen, already existed near the shore of the Bangka Strait with 117 resident families. The villagers used the more fertile areas of the site for agriculture and gardens. The land was public land, and the district government and the hotel owner agreed on an amount of compensation the villagers would be paid for the loss of the use of the land. Forty-two families chose to receive a new house and ownership of the land on which it stands, in a new location at the edge of the project site, approximately one kilometre from the shore and one kilometre from Maen in the opposite direction. The other families chose financial compensation.

As well, many families from the main part of the village of Maen had gardens on the project site where they raised vegetables, coconuts and domestic animals. They were also awarded some financial compensation for the loss of the use of the land

The Paradise AMDAL

Paradise began to acquire land for the project in 1990, and was subjected to a full environmental impact assessment that began in 1991 (Massie 1997). The Paradise ANDAL is a thick document, consisting of approximately 270 pages, including several maps and figures, plus another fifty pages of attachments. It is divided into eight sections with the following headings: introduction, activities planned, profile of the original environment at the site, predicting impacts, impact evaluation, mitigating impacts, directions and alternatives for the Environmental Management Plan and the Environmental Monitoring Plan, and conclusion. The attachments include permits and approvals for the hotel from a variety of government institutions, as well as biographical data on the AMDAL team that performed the study.

The RKL and RPL (the environmental management and monitoring plans) are both much shorter, each with approximately eighty pages divided into two sections, plus around fifteen pages of attachments. The first section describes the importance of environmental management or monitoring respectively, and briefly lists the environmental components of the Paradise project that will be managed and monitored. Several maps of the location of impacts are included. The second section is a detailed list of all management and monitoring activities to be performed, listed according to project phase. The attachments include matrices that show the impacts to be managed or monitored and the parties involved in the process, and flowcharts that show the connections between impacts.

The impacts of Paradise

Physical environment

The ANDAL characterizes the area as not particularly fertile, but of great aesthetic beauty due to the clear water, healthy coral, dramatic rock structures, rolling hills and waving coconut palms. The terrestrial ecosystem is considered to be healthy and stable. It was determined that the sea life in the area is not highly diverse, but normal for the region. The RKL for the site suggests that local fishermen will be allowed to continue to work, but in a smaller area, and their use of chemicals will be prohibited. They will also be instructed not to take certain species of fish.

According to the project ANDAL, in preparing the land for building, some small hills were flattened and swamps were filled, a few small mangrove trees were removed, and many coconut palms were cut down and then used to build the new houses for the residents of Tamba. Most building supplies were brought in from other areas, and the government improved the road to the area at this time as part of a project to develop the area for tourism. As well, the hotel paved the old route from the main road to the hotel, which passes through part of Maen. Since the land was previously used for agriculture, no virgin wilderness was destroyed and few highly-valued species of wildlife inhabited the region.

The main water source for the hotel is the river Maen, and the site's waste water is emptied into another river, the Pangisang. The liquid waste first goes through a sewage treatment process, so it does not greatly increase pollution in the river. In addition, locals are now restricted from bathing or washing clothes in the rivers. The hotel has a landfill on site for solid waste, and attempts to prevent littering of the area include regular waste collection.

The ANDAL predicts that potential significant impacts on the physical environment during the construction phase would include: an increase in noise levels, land levelling, sea-water pollution, and a decrease in biodiversity of terrestrial vegetation on site. The RKL deals with each of these impacts, with activities aimed at mitigating all except the levelling of the land, which is seen to be a positive impact. Although the RKL was not actually complete at the time of construction, interviews with hotel staff and community residents indicate that these impacts were well managed at the time. Only a minimum of trees were cut down and water samples collected near the end of the construction phase showed that the quality had not deteriorated significantly from its original level (Massie 1997).

Each of the negative potential impacts has avoidance or mitigation techniques incorporated into the management plan for the project, and those that are continual are also scheduled for monitoring in the RPL. The ANDAL also suggests that a jetty with its diving operation should be relocated. However, no alternative location is offered and the suggestion is not considered essential to environmental management of the site, and is not included in the RKL.

One component of the physical environment that is not considered in the AMDAL is shoreline erosion but this has yet to be a problem. Another, more significant omission in the AMDAL is that not all of the impacts of the diving operation are considered, including those that affect the Bunaken Marine Park. While the potential for localized water pollution near the pier is discussed, impacts on coral reefs further offshore and in the park are not.

In the absence of current soil, vegetation, air or water quality samples, to all appearances the Paradise resort is working to both protect and enhance the living environment in the area. One of the resort managers interviewed stated that the environment is the object of tourism there, so they simply cannot afford to destroy it. However, careful environmental monitoring is essential to ensure that negative changes do not emerge in the future.

Social environment

The project's social impacts on local communities have been both more widespread and less positive than the physical changes. The ANDAL predicted that the Paradise project would have almost no negative social impacts aside from a slight lessening of local support for the project due to increased noise and the influx of workers from other areas, while forecasting a large number of very positive social and economic impacts on the residents of Tamba and Maen. Predicted positive changes included: increasing local incomes, improved satisfaction with work, increased education levels, greater business opportunities, better housing, satisfaction with owning homes and land; and improved infrastructure in the district. While a few of these expected benefits have been realized, most of them have not and a number of negative impacts have emerged in their stead. Some of these negative impacts were not predicted at all, whereas others were forecast and methods of mitigating or minimizing them are outlined in the RKL, but have not been carried out. Management actions aimed at ensuring that some of the positive impacts are optimized are also outlined in the RKL, but have not been performed.

The AMDAL team assessed the social impacts of the Paradise project during the pre-construction stage, after the residents of Tamba had been told what their compensation would be for their forced relocation. Interviews held with villagers from Tamba revealed that the amount of money offered as compensation for the land was widely regarded as insufficient (Pt Dinnelator Likupang Beach Paradise 1997). While the ANDAL does not state the amount given, interviews revealed that it was very low, at 35 Rupiahs/m^2, or 1–2 cents/m^2 in Canadian terms. Many families from Maen also had gardens on the project site where they raised vegetables, coconuts and domestic animals, and they were awarded similar financial compensation for the loss of the use of the land. A frequently heard observation was that the money was not enough to buy a pack of matches – they cost 50 Rupiahs. Some of those concerned with the project are of the opinion that the amount of compensation was reasonable but that the villagers squandered it unwisely (Massie 1997); however, the general consensus is that the rate of financial compensation was very low. Although the ANDAL does mention the issue, and suggests that the villagers' dissatisfaction could have a negative impact on the community's perception of the project, since the amount was determined by the 'Tripeka', which is the triad of government, military, and police, and agreed to by the hotel owner, it was non-negotiable. In other parts of the ANDAL, the payment of compensation is listed as having the significant positive impact of increasing the community's income.

Other than expressing dissatisfaction with the compensation for the land they depended on, the residents of Tamba were strongly supportive of the project when the AMDAL team interviewed them. The government assured the local communities that the first priority in tourism development would be to use local workers in all phases of the project. It was envisioned that there would be a swing from agriculture to tourism as the major regional industry, with the children of farmers and fishermen turning to tourism for employment. At the time of the AMDAL survey, 100 per cent of respondents voiced support for the hotel, as they felt it would open new opportunities for work and enterprise. Many young people were very enthusiastic about working in the tourism industry. They hoped to receive training at the hotel, both in hospitality jobs and in foreign languages. Older residents foresaw available work during construction, and the potential to sell food and other goods to workers, hotel staff and tourists.

Local employment realities have not matched local expectations. While most of the community was employed by the hotel during construction, few local people are currently employed there. Of the approximately 250 hotel employees, all are Indonesian and about 95 per cent are from Sulawesi, most coming from North Sulawesi, but very few are actually from the Likupang district. Conflicting reports of numbers exist. Officially, as many positions are filled by locals as possible: the project management plan states that more than 80 per cent of the workforce must be local. One manager claims that the workforce is approximately 50 per cent local and other staff at the hotel say there are very few locals working there. The village headmen of Tamba, Maen and Wineru all report that only between five and ten residents of each of their villages work at the hotel. The majority that do work there are in menial cleaning or gardening positions. They receive no training and the wages are low. As well, they are paid daily, meaning that if they are sick and miss work, they have no income. While the rate of pay is above the legislated minimum, the local people suggest that it is not enough to support a family. Two respondents reported that they had to quit working at the hotel because the income was insufficient to meet their needs.

In support of the expected benefits to local residents, the RKL, which was agreed to by the hotel owner and approved by regional government authorities, suggests that the hotel owner will provide residents of both Tamba and Maen with training in the hospitality industry, and courses in cottage industries based mostly on producing local crafts for sale in the hotel gift shop. The RKL also suggests that the owner will offer seminars to teach local people entrepreneurial activities. So far, there is little evidence of any such efforts. The courses and seminars have not been offered, no souvenirs or local handicrafts are available in the town, and the gift shop at the hotel is very small and poorly stocked. No locals work in white-collar positions at the hotel, and no in-house training is offered for service industry positions. A few residents report that members of the local community are not allowed onto the project site, so there is no possibility of selling either produce or handicrafts directly to the hotel. The hotel does buy fish and produce at the local market, in the district's main village, but many residents of Tamba and Maen pointed out that they no longer have land to grow vegetables or coconuts to sell. While some traditional dancers from the area do perform at the hotel, they are from the slightly more distant villages of Wineru and Winuri.

The ANDAL, RKL and RPL documents for Paradise were not finalized until March of 1997. Since the hotel first opened for business in December of 1995, it had already been operating for over a year before the management plan was completed. Clearly, many elements of the RKL could not have been implemented during earlier phases of the project, and there is little evidence to suggest that the hotel owner has any intention of fulfilling the requirements outlined in it for the operational phase either. The heads of each village are not aware of any plans for cottage industry courses, and neither is in contact with either the hotel owner or with the BLH, the government agency then responsible for implementing environmental management plans in the province.

In the ANDAL it is predicted that the community's income would increase as work became available during the construction and operational phases of the project and entrepreneurial opportunities to sell goods and materials to the hotel opened up. While some local people have increased their income level, in defining increased income as a significant positive impact, the ANDAL ignores the fact that when the project site land was available for agriculture, the residents had less need for money as they did not have to buy food. In fact, the ANDAL does not deal effectively overall with the issue of land use and the impacts incurred by the community when they lost their farmland. Although the ANDAL includes study of an environmental component entitled 'Patterns of Land Ownership and Land Use', the emphasis is on ownership, and the impact is determined to be positive. The only factor considered is that the resettled villagers received ownership certificates for their new houses and the land on which they stand. In fact, land ownership and land use are very different things and in this case, while it is true that the residents of Tamba are pleased at owning their new homes, they are extremely dissatisfied with the loss of land for agriculture and the insufficient compensation. Ideally the two issues require separate treatment, and the AMDAL should acknowledge the negative impact experienced by the villagers and the potential follow-on impact on their attitude toward the project. If the promised hotel work had appeared, perhaps the income would have replaced the land loss. As it is, this impact has led to some disintegration of the community as many people have moved because they could not earn enough there to support their families.

As well as the potential for work at the hotel, the people from Tamba expected better living conditions in the new location, with healthier houses, electricity, clean water and a good road. The houses in the new village vary in size, and many are in different stages of renovation or having additions added. They are almost all larger and sturdier in appearance than most of the houses in Maen, although several Tamba residents say that they were smaller to start with and are being expanded because the members of the village participate in '*arisan*', a traditional practice where each family contributes a set amount of money to a communal kitty which is given to a different family each month for help with construction. Piped water is available to villagers, however it is expensive and very few of them use it, the majority continuing to use well water. Now, although there have been many positive changes, the people of both parts of Tamba are very dissatisfied with the current situation, and many voiced strong disapproval of the Paradise Beach Project. They all feel that the village was promised prosperity with the development of tourism, but that promise was not fulfilled and they are now poorer than before. The villagers in general feel that they are unable to do anything about their situation, and express an attitude called '*pasrah*', which embodies the concept of submitting to fate.

Moving slightly further away to the main part of the village of Maen, approximately half of residents interviewed previously had gardens on the land that was taken over by the project and they were all very dissatisfied with the financial compensation they received. All spoke of the difficulty of earning money to buy the food that they formerly grew themselves, and many of them felt that the amount they had been paid was so little that it was nothing more than a friendly gesture on the part of the hotel owner.

Unlike the very unhappy people of Tamba though, less than half expressed a negative attitude toward the hotel and many others expressed ambivalence, saying the hotel had influenced their lives very little if at all. The people of Maen are predominantly of the Islamic faith which places great importance on modest dress, and several of the members of the village expressed disapproval of the bathing costumes that tourists often wear. However, this disapproval was only in one case strong enough to escalate into disapproval of the entire tourism development. One, a civil engineer who works for the hotel, views the project as having a strong positive impact on the community.

The statistical information in the ANDAL report is not always apposite. For example although the ANDAL contains a great deal of demographic and social information, little of it is analysed to determine the elements that are vulnerable to impacts or the relationships between environmental components and potential impacts. For example, no connection is drawn between the education levels and the prospects for employment in the tourism industry. No indication is given of the relationship between religion or education level and the impacts of new influences. In summary, while there is an abundance of information on social aspects of the region, it is not always evaluated usefully.

Strengths and weaknesses of the Paradise AMDAL process

Indonesia has sophisticated regulations requiring the conduct of detailed environmental impact assessments for development projects. Nevertheless, the AMDAL process has been less effective than it should be at the Paradise resort. Many of the social impacts of the project have been poorly assessed and poorly managed. Many of the failings of the AMDAL process in this case are a result of weak environmental policy implementation, while others are a result of insufficient guidelines for the assessment of tourism developments, or insufficient training in SIA.

The Paradise ANDAL is a very long and thorough document. It provides a comprehensive profile of the original physical and social environments around the project site. Explanations of the planned activities are detailed, and the impacts they are likely to cause are clearly indicated, as are the environmental components that are vulnerable to impacts. The collection of data is thorough and expert. All physical and biological samples were taken in two different years in both the dry and wet seasons, and therefore have high validity. The ANDAL clearly explains the criteria by which each impact is determined to be significant or not significant, thus justifying those that are then implicated for managing and monitoring. The criteria used are those listed in published guidelines for determining the significance of impacts.

However, the fact that the AMDAL process was not completed until so late in the project cycle makes it evident that it was not considered as part of the feasibility study. The timing of the AMDAL documentation also impeded the possibility of public participation in the project cycle. The Paradise project is an example of a situation in which no public participation whatsoever occurred. Since the Paradise ANDAL, RKL and RPL were not approved until long after the operational phase of the hotel began, and no copies of the documents were made available to the villagers in advance, there was no forum for them to participate in. The majority of those interviewed had no idea that they were entitled to make their views known to the government and project initiators; and those who said that they were aware of their legal right insisted that no opportunity had existed for them to take part in any stage of the decision-making process. In fact, most of those interviewed expressed the feeling that the changes had been imposed on them with no local say in the matter whatsoever.

The AMDAL team did not predict the very significant impacts resulting from the residents of Tamba losing the use of land for agriculture. This situation reflects the general trend toward poor social impact assessment within AMDAL, and likely is a result of a combination of factors. The format of the report on social impacts follows Depparpostel's guidelines for AMDAL for tourism developments very closely, a fact that emphasizes the need to improve these guidelines in the area of SIA. It is also likely that the members of the AMDAL team did not have sufficient training and experience in social impact assessment.

The RKL outlines management activities that could mitigate many of the very negative social impacts that have emerged, but the RKL has not been implemented. Although the project's physical impacts seem to be well managed, they are not being monitored, allowing for the very real possibility of negative change in the future. The government's lack of commitment to enforcement of environmental policy is a major factor behind the situation. Confusion between the various responsibilities of the agencies involved in AMDAL is very likely another cause of inaction in the Paradise situation. It is also unclear what party has enforcement powers. Also, the hotel was already operating before the AMDAL was approved, so clearly the policy of linking permits to implementation of the management and monitoring plans was not enforced, nor has the hotel owner been fined for failing to follow the two plans.

Although the monitoring plan is also approved and theoretically functioning, there is no indication that any monitoring is actually taking place. The head of the AMDAL team that did the original study is not aware of any activities, and interviews with the village headmen of Maen and Tamba reveal that no monitoring team has visited the area. The Environmental Studies Centre at Sam Ratulangi University in Manado is listed in the RPL as responsible for post-construction monitoring of noise, water quality and land vegetation, by taking a variety of samples at pre-determined time intervals. This had not been initiated as of September 1997.

CONCLUSIONS

Although tourism does have negative impacts on the environment, and tourists do use resources, it is an industry well positioned to contribute to sustainable development. Haywood (1993) points out that many interest groups, from tour agents to destination communities to individual tourists, are aware that the tourism experience suffers due to environmental degradation, crowding and pollution. The tourism industry, in turn, is often reliant on high-quality environments and it would be in the industry's own best long-term interests to control its growth to avoid these impacts. Hunter (1995b) suggests that sustainable tourism development can occur when it is understood that the long-term economic viability of tourism projects depends on controlled growth and successful, sustainable management of the resources on which they are based. Mieczkowski (1995: 376) suggests that, when it is regulated, tourism is 'one of the most environment-friendly economic development options to choose from, and is, therefore, likely to expand and increase its share in the world's GNP'.

The requirement that developers undertake thorough and timely impact assessments and act upon their results is one such form of regulation. Tourism development can have far-reaching consequences for destination area communities and EIA is potentially a valuable tool for environmental planning – it can be used to formulate national policies on development, to determine the best alternatives among a variety of proposed projects, to determine the best development alternatives within an approved project, to minimize negative impacts of development, and to find the optimal level of positive impacts. EIA can be used either at the regional planning level or, as discussed in the case study above, at a site-specific level, and could play an important role in promoting sustainable development. However, it is an undertaking fraught with challenges, particularly when applied to tourism.

Indonesia has a legislated EIA process, AMDAL, that includes assessment of the social impacts as well as physical impacts of development projects. To date, AMDAL has been used almost exclusively on a project-specific basis rather than at the broader planning level. AMDAL is mandatory for all development projects expected to have significant impact on the environment, and the government has also compiled a list of sensitive locations and ecosystems that automatically require EIA before a development can proceed (Dick and Bailey 1994). However, little mention is made of tourism developments in the country's AMDAL literature, despite the fact that tourism has been growing rapidly and the government is encouraging its expansion. AMDALs are frequently completed too late to be a part of project feasibility studies and, in fact, the process often does not begin until well after the project is approved and may not be completed until construction has already begun. In such a situation, AMDAL findings and recommendations cannot be considered in the decision-making process, local input is precluded and environmental and social concerns are neglected: developments proceed in inappropriate locations with few environmental checks. However, achievement of the full potential of impact assessment requires commitment of resources, skilled personnel and political will, and these are all in short supply in Indonesia. Although there is growing recognition of the importance of tourism, Indonesia is not currently well prepared to assess, monitor or manage the impacts of tourism.

Indonesia has not been singled out for criticism. In fact, in many ways its legal and administrative frameworks are a model for others to follow. But, unfortunately, there is a wide gap between theory and practice. Yet the rapid growth of tourism in Indonesia and many other developing nations suggests a need to develop assessment methods specifically for tourism in these countries. However, the characteristics of tourism make this a challenging undertaking.

Hunter's (1995a) suggestions for EIA use in tourism developments include: studying social and economic as well as environmental impacts; encouraging public consultation; using selection criteria that allow for screening, so that only those developments likely to have significant environmental effects will be studied; using a scoping procedure, so that resources will not be wasted on studies of insignificant impacts; attention to and careful assessment of cumulative impacts; recognition and assessment of the variation of impacts due to seasonal patterns; and determination of the size of the area around proposed tourism developments that should be considered, which will depend on the local situation. These are wise suggestions which have global applicability, particularly as tourism can expose environments and communities suddenly to massive change, new technologies, new people and new ways of life.

Acknowledgments

The field research reported in this paper was funded in part by a grant from the Social Sciences and Humanities Research Council of Canada to G. Wall and by the University Consortion on the Environment through the Canadian International Development Agency. Research was facilitated by Dr Louise Waworunto and Mr Denny Karwur of Universitas Sam Ratulangi, Manado and by the Environmental Studies Centre's Development in Indonesia project. The important contributions of Ibu Maasje Massie, Ibu Fransine Manginsela and Ibu Julie Waworuntu are gratefully acknowledged.

References

BAPEDAL (1996) *Himpunan Peraturan Tentang Pengendalian Dampak Lingkungan*, Seri II, Jakarta, Indonesia: BAPEDAL.

Biswas, A.K. (1992a) 'Preface', pp. vii–ix in A.K. Biswas and S.B.C. Agarwal (eds) *Environmental Impact Assessment for Developing Countries*, Oxford: Butterworth–Heinemann.

—— (1992b) 'Summary and Recommendations', pp. 237–45 in A.K. Biswas and S.B.C. Agarwal (eds.) *Environmental Impact Assessment for Developing Countries*, Oxford: Butterworth–Heinemann.

Butler, R.W. (1993) 'Pre- and post-impact assessment of tourism development', pp. 135–55 in D.G. Pearce and R.W. Butler (eds) *Tourism Research: Critiques and Challenges*, London: Routledge.

Carter, J.A. (1997) *Assessment of Coastal and Marine Resource Issues in Sulawesi and Analysis of Related Institutional Needs of BAPEDAL Wilayah III*, Ujung Pandang: Collaborative Environmental Project in Indonesia (CEPI) and the Canadian International Development Agency (CIDA).

Ceballos-Lascuráin, H. (1996) *Tourism, Ecotourism, and Protected Areas*, Gland, Switzerland: IUCN.

Conover, S.A.M. and Hanson, A.J. (1992) *The Development of Environmental Assessment (AMDAL) in Indonesia*, Environmental Management Development in Indonesia Project, Environmental Reports 22, Halifax, Canada: Dalhousie University.

Dick, J. and Bailey, L. (1994) *Indonesia's Environmental Assessment Process (AMDAL): Progress, Problems and a Blueprint for Improvement*, Environmental Management Development in Indonesia Project, Environmental Reports 29, Halifax, Canada: Dalhousie University.

Directorate General of Tourism and United Nations Development Program (1992) *Tourism Sector Programming and Policy Development, Output 1. National Tourism Strategy*, Jakarta: Government of Indonesia.

Doberstein, B. (1992) *An Evaluation of the Use of Environmental Impact Assessment for Urban Solid Waste Management in Denpasar, Bali*, Student Paper Number 12, Waterloo, Canada: University Consortium on the Environment Publication Series, University of Waterloo.

Hardin, G. (1968) 'The tragedy of the commons', *Science* 162: 1243–8.

Haywood, K.M. (1993) 'Sustainable development for tourism: a commentary with an organizational perspective', pp. 233–41 in J.G. Nelson, R. Butler, and G. Wall (eds) *Tourism and Sustainable Development: Monitoring, Planning, Managing*, Heritage Resources Centre Joint Publication Number 1, Waterloo, Canada: Department of Geography, University of Waterloo.

Hunter, C. (1995a) 'Environmental Impact Assessment and Tourism Development', pp. 122–68 in C. Hunter and H. Green (eds) *Tourism and the Environment: A Sustainable Relationship*, London: Routledge.

Hunter, C. (1995b) 'Key concepts for tourism and the environment', pp. 52–91 in C. Hunter and H. Green (eds) *Tourism and the Environment: A Sustainable Relationship*, London: Routledge.

Hunter, C. and Green, H. (1995) (eds) *Tourism and the Environment: A Sustainable Relationship*, London: Routledge.

Kinnaird, M.F. and O'Brien, T. (1996) *Tangkoko–Duasudara Nature Reserve, North Sulawesi Draft Management Plan*, A Wildlife Conservation Report prepared for the Directorate of Nature Conservation, Directorate – General of Forestry, Republic of Indonesia.

Knight, D., Mitchell, B., and Wall, G. (1997) 'Bali: sustainable development, tourism and coastal management', *Ambio* 26 (2): 90–6.

Long, V. (1992) 'Social mitigation of tourism development impacts: Bahias de Huatalco, Oaxaca, Mexica', in C.A.M. Fleischer-van Rooijen (ed.) *Spatial Implications of Tourism*, Groningen: Geo Pers.

Martopo, S. and Mitchell, B. (eds) (1995) *Bali: Balancing Environment, Economy and Culture*, Department of Geography Publication Series 44, Waterloo: University of Waterloo.

Massie, M. (1997) Personal communication with AMDAL team chairperson. Written notes, July 1997.

Mathieson, A. and Wall, G. (1982) *Tourism: Economic, Physical and Social Impacts*, London: Longman.

Mieczkowski, Z. (1995) *Environmental Issues of Tourism and Recreation*, Maryland: University Press of America.

Ministry of State for Environment (1996) *Indonesia's Marine Environment: A Summary of Policies, Strategies, Actions and Issues*, Jakarta.

Pearce, D. (1989) *Tourist Development*, 2nd edn, London: Longman.

Pt Dinnelator Likupang Beach Paradise (1997) *Analisis Dampak Lingkungan (ANDAL), Kawasan Pariwisata, Dinnelator Likupang Beach Paradise and Recreation Facilities*, Manado: Paradise Hotel and Resort.

Richter, L. (1993) 'Tourism policy-making in South-East Asia', pp. 179–99 in M. Hitchcock, V.T. King, and M.J.G. Parnwell (eds) *Tourism in South-East Asia*, New York: Routledge.

Roberts, P. and Hunter, C. (1992) 'Environmental assessment: taking stock', *Working Paper* 11, Leeds: Centre for Urban Development and Environmental Management, Leeds Metropolitan University.

Shera, W. and Matsuoka, J. (1992) 'Evaluating the impact of resort development on an Hawaiian Island: implications for social impact assessment policy and procedures', *Environmental Impact Assessment Review* 12 (3): 349–62.

Smith, S.L.J. (1988) 'Defining tourism: a supply side view', *Annals of Tourism Research* 15 (2): 179–90.

Sofield, T.H.B. (1995) 'Indonesia's national tourism development plan', *Annals of Tourism Research*, 22 (3): 690–3.

Telfer, D.J. (1996) *Development Through Economic Linkages: Tourism and Agriculture in Indonesia*, PhD Thesis, Waterloo: Department of Geography, University of Waterloo.

Wall, G. (1996) 'Rethinking impacts of tourism', *Progress in Tourism and Hospitality Research* 2 (3/4): 207–15.

—— (1997) 'Indonesia: the impact of regionalization', pp. 138–49 in F. Go and C. Jenkins (eds) *Tourism and Development in Asia and Australasia*, London: Cassell.

Werner, G. (1992) 'Environmental impact assessment in Asia: lessons from the past decade', pp. 16–21 in A.K. Biswas and S.B.C. Agarwal (eds) *Environmental Impact Assessment for Developing Countries*, Oxford: Butterworth–Heinemann.

Whitton, A.J., Mustafa, M., and Henderson, G.S. (1987) *The Ecology of Sulawesi*, Yogyakarta: Gadjah Mada University Press.

World Bank (1994) *Indonesia: Environment and Development*, A World Bank Country Study, Washington DC: The World Bank.

15

IMPACT OF KOREAN TOURISTS ON KOREAN RESIDENTS IN HAWAII AND QUEENSLAND, AUSTRALIA

Sang Mu Kim

INTRODUCTION

Travel abroad by Koreans in large numbers is a recent phenomenon. Prior to 1989 the Korean Government prevented its citizens travelling overseas in large numbers by imposing restrictions that included age and travel duration limitations, foreign exchange controls, purpose of trip controls and requirements to lodge large deposits of money (1983, over US$2,500) with government agencies. The aim of this policy was to build up Korean currency reserves. In tandem with this policy, the government fostered domestic tourism by heavily investing in the domestic tourism infrastructure. This policy inadvertently created a taste for tourism that was ultimately to translate into a desire for overseas travel.

As Korea's balance of payments situation improved and the demand for overseas travel by its citizens increased, restrictions gradually became more flexible and were lifted completely on 1 January 1989. In the first year of full liberalization, outbound departures increased by 67.3 per cent to 1,213,112 persons. During that year the growth in pleasure tourists versus other categories was particularly notable, with an increase of 235.2 per cent over the 1988 total, and accounted for 37.0 per cent of the total outbound market (KNTO: Korea National Tourism Organization 1990).

The subsequent growth in outbound travel was rapid, increasing by 526.3 per cent in the seven-year period 1989 to 1995. In 1995, the number of Korean outbound travellers reached 3,818,740, which outnumbered inbound foreign visitors for the first time. The growth rate was 21.1 per cent over the previous year (KNTO 1995). In terms of destinations, the 1995 Korean Tourism Annual Report indicated that Japan (26.2 per cent) was the most popular destination, followed by the United States of America (17.2 per cent), China (10.6 per cent), Thailand (8.0 per cent) and Hong Kong (5.9 per cent). By continent, Asia accounted for 64.1 per cent of Korea's total outbound market in 1995, followed by the Americas (19.2 per cent), Europe (8.0 per cent), Oceania (8.0 per cent) and Africa (0.7 per cent).

However, such a rapid increase in the number of Korean outbound tourists caused some problems. For example, Korea's tourism balance of payments went into the red in 1991, recording a deficit of US$358 million, which grew to US$523 million in 1992. In 1992, the balance turned positive due to the exclusion of expenses spent by overseas students, but in 1994 and 1995 it again showed deficits of US$282 million and US$494 million respectively (KNTO 1996; Table 15.1). Furthermore, a negative image of Korean tourists had developed among residents in some tourist destinations because of the different culture and customs of the new tourists (Kim 1993).

The problems mentioned above need to be considered not only from the economic point of view but should also be seriously examined from the sociocultural perspective. It has long been noted that tourism can be an important mechanism for increasing international understanding (Mathieson and Wall 1986). This process is understood to be generated from the process of mutual contact between tourists and host. Obviously, employees in the tourism industry have direct contact with tourists. Analysing the relationship these employees have with tourists could enable us to measure the economic and sociocultural impacts caused by international tourists and

Table 15.1 Tourist arrivals and balance of payments (1989–1995)

Year	*Visitor arrivals*		*Korean departures*		*Tourist receipts*		*Korean travel expenditures*		*Balance*
	000s	*%**	*000s*	*%**	*US$ million*	*%**	*US$ million*	*%**	*US$ million*
1989	2,728	(16.6)	1,213	(67.3)	3,556	(8.9)	2,602	(92.2)	955
1990	2,959	(8.5)	1,561	(28.7)	3,559	(0.1)	3,166	(21.7)	393
1991	3,196	(8.0)	1,856	(18.9)	3,426	(−3.7)	3,784	(19.5)	−358
1992	3,231	(1.1)	2,043	(10.1)	3,272	(−4.5)	3,794	(0.3)	−523
1993	3,331	(3.1)	2,420	(18.4)	3,475	(6.2)	3,259	(−14.5)	216
1994	3,580	(7.5)	3,154	(30.3)	3,806	(9.5)	4,088	(25.4)	−282
1995	3,753	(4.8)	3,819	(21.1)	5,587	(46.8)	5,903	(44.4)	−316

Source: Korea National Tourism Organization 1996.
*% change over previous year.

allow us to understand better the repercussions which will affect the images of Korea and its people as a whole.

According to KNTO's survey, 50.8 per cent of Korean overseas travellers responded that their main reason for travel was to 'experience overseas travel', followed by 'to visit a particular city or place' (42.3 per cent), and 'reasonable price' (24.5 per cent). This indicates that most Koreans considered the purpose of overseas travel as 'for a change' or 'to improve their social status' (Table 15.2). Respondents were asked to complete these questions in a descriptive manner. However, the survey is less likely to be useful for assessing the actual impacts since the survey respondents themselves are subjects to be judged, and perhaps not appropriate also to be the judges. Therefore, it would be more appropriate and objective to survey residents of destinations, particularly tourist business proprietors in destinations, in order to more objectively identify sociocultural impacts as well as economic impacts.

The initial years of Korean outbound tourism have seen many Korean tourists travelling initially on escorted tours and in groups to reduce problems of language and customs in unfamiliar settings, as did Japanese tourists in their initial period of growth. Along with this, Korean entrepreneurs began to operate a range of businesses in destinations popular with Korean tourists, seizing an opportunity for economic development and also providing a perceived needed range of services in the Korean language and customs. Thus, in some destinations Korean tourists can stay in Korean-operated accommodation, shop in Korean shops, eat Korean food and use Korean guides. Other nationalities undertake similar actions in other locations (the British in Mediterranean destinations being a notable example). The Korean businessmen engaged in such activities are in a unique position to serve their compatriots, assisting the interaction between tourists and residents and being able to act as cultural and economic brokers. Their role and perceptions have been little researched to date, and this study is somewhat of a pilot effort to draw upon the reactions to Korean tourists and their effects on destinations.

Table 15.2 Reasons to travel overseas

Category	*Rank (%)*
To experience overseas travel	1 (50.8)
To visit a particular city or place	2 (42.3)
Reasonable price	3 (24.5)
No interesting place to visit in the country	4 (12.8)
To obtain new information and knowledge	5 (7.2)
To appreciate overseas custom and culture	6 (1.5)

Source: KNTO 1995.

PURPOSE OF STUDY

Research on the economic effects of tourism has a much longer history than that on less tangible impacts. A considerable amount of research has been undertaken on economic impacts (Archer 1973, 1977, 1982; Sadler, Archer and Owens 1973; Archer and Wanhill 1981, 1985; Wanhill 1983, 1988; Fletcher 1987), but less specific research is found on sociocultural impacts (Belisle and Hoy 1980; Var, Kendall and Tarakcioglu 1985; Ap 1990). Tourism can have a significant impact upon residents in a destination country, however, and upon tourist business proprietors in particular. Inbound tourism usually has a positive effect on the economy in terms of foreign exchange, income and employment (Mill and Morrison 1985), while some destinations have experienced negative sociocultural effects (Mathieson and Wall 1982; Kim 1993).

Korea has two points at issue to be resolved regarding outbound tourism: first, a deficit in the balance of international tourism payments since 1991, and second, the negative sociocultural impacts of Korean tourists on residents of destinations, including Korean residents in these locations. These issues have been particularly criticized by newspaper reporters, who have recorded a negative image of Korean outbound tourism throughout the country (Kim 1996). Some Koreans even feel guilty to be an overseas tourist.

It is dangerous to publicize these issues without any proper evidence or proof based on scientific research because it may create an impediment to Korean outbound tourism as well as creating an inaccurate image of Korea and Koreans. In consequence, these issues need to be studied and explored properly so as to identify whether the residents in a destination derive real economic benefits from Korean tourists. Simultaneously, the sociocultural impact of Korean tourists on the residents must be examined to identify clearly and measure any effects.

Following the relaxation of travel restrictions on Korean citizens in 1989, Korean outbound tourism to Hawaii and Australia, among other destinations, has been growing remarkably. In 1995, Korean tourists to Hawaii and Australia numbered 104,550 and 167,975 respectively, which made Koreans the fifth largest market for both destinations (KNTO 1995). In response to increasing Korean overseas tourism to these destinations, some Koreans who had little experience in tourism management moved to these destinations to operate tourism businesses.

The purpose of this chapter is to explore the economic and sociocultural impacts of Korean overseas tourists on Korean residents who are involved in tourist businesses in Hawaii and Queensland, Australia, so as to identify positive and negative impacts on the residents. Based on the results of the study some recommendations and appropriate strategies to improve the situations will be suggested.

SURVEY ADMINISTRATION

Hawaii has been known to Koreans as a 'paradise on earth' for a long time and attracted 104,550 Korean tourists in 1995. This figure represented a 272.8 per cent growth rate over a seven-year period (38,330 Korean tourists in 1989). On the other hand, Australia is an emerging and popular new tourist destination for Koreans. It attracted 167,975 Korean tourists in 1995, showing a 6.4 per cent growth rate over the same seven years (10,406 Korean tourists in 1989). According to the assessment of Korean tourist businessmen in Hawaii and Queensland, Australia, the most attractive features for Korean tourists in both destinations are beautiful scenery, followed by clean beaches, tourist activities and kind and friendly people (Table 15.3).

The data for this study were compiled from two groups; fifty-two Korean tourist business proprietors residing in Hawaii and forty Korean residents engaged in tourist businesses in Queensland. The information on the business populations was obtained from the Korean Tourism Agents Association of Hawaii in Honolulu and Korean Tourists Association of Queensland on the Gold Coast. In order to collect the data, survey interviews with seventy Korean tourist businessmen in Hawaii were conducted by the author during the period of 1 January to 28 February 1993; fifty-one Korean tourist businessmen in Queensland were interviewed during the period of 1 March to 30 April 1996. The subject of the surveys included registered tourist businesses on the lists of both associations. They are broken down into three categories: travel agencies, restaurants and gift shops. The response rates were 74.3 per cent in Hawaii and 78.4 per cent in Queensland (Table 15.4).

Data were collected on Korean tourist businessmen's revenue, number of employees, popular shopping items, the perceived positive and negative impacts of Korean tourists and residents' attitude towards Korean tourists.

Table 15.3 Attractive features in the destination

	Hawaii		*Queensland*	
Category	*Rank*	*%*	*Rank*	*%*
Beautiful scenery	1	(96.2)	1	(97.4)
Clean beaches & environment	2	(92.9)	2	(94.2)
Tourist activities	3	(59.5)	4	(66.6)
Kind and friendly people	4	(51.2)	3	(72.4)
Pleasant climate	5	(49.2)	5	(45.8)
Traditional culture & customs	6	(26.1)	6	(30.2)
Good shopping opportunity	7	(6.5)	7	(8.4)
Unique cuisine	8	(3.5)	8	(6.6)

Source: Survey data.

The number of data were not sufficient enough to run Chi-square or ANOVA (Analysis of Variance) tests with the Statistical Analysis System (SAS) program. However, the observations were used to derive frequencies and percentages based on simple statistical analysis.

ANALYSIS AND FINDINGS

Economic impact

According to KNTO's survey, the per capita expenditure of Koreans travelling overseas in 1994 was US$2,011, which showed a 24.8 per cent increase over the previous year. The average expenditure on accommodation per Korean tourist was US$600 or 29.8 per cent of total expenditures, followed by shopping at US$412 (20.5 per cent), food and beverage at US$312 (15.5 per cent), and entertainment at US$262 (13.0 per cent). Unidentified or other expenditures per Korean tourist were US$425 or 21.2 per cent of total expenditures (Table 15.5).

However, the average expenditure per Korean tourist to Hawaii was US$1,637, which is 18.6 per cent less than the average of total Korean overseas travellers' expenditures. The main reason for this is that the average length of stay in Hawaii is only 4.9 days while the total average length of stay of Korean overseas travellers is 11.2 days. Although the average length of stay in Australia is 7.2 days, which also is less than the total average length of stay, the average expenditure per Korean tourist to Australia was US$2,230, which is 10.9 per cent greater than the overall average. This is mainly because Korean tourists to Australia tend to spend more money than the average on purchasing local products such as health food, opals and sheepskin goods.

The average annual sale revenue for Korean tourist businessmen in Hawaii was US$875,000, and 74.6 per cent of this total was earned from Korean

Table 15.4 Status of the respondents classified by category

Category		*No. of subjects*	*Respondents*	*Response rate %*
Travel agency	Hawaii	30	24	80.0
	Queensland	17	16	99.1
Restaurant	Hawaii	25	18	72.0
	Queensland	18	13	72.2
Gift shop	Hawaii	15	10	66.7
	Queensland	16	11	68.8
Total	Hawaii	70	52	74.3
	Queensland	51	40	78.4

Source: Survey data.

Table 15.5 Koreans' overseas travel expenditures by category

Category	*1992 US$*	*1993 US$*	*1994 US$*
Accommodation	550	537	600
Food & beverage	287	275	312
Entertainment	265	250	262
Shopping	425	312	412
Others	362	237	425
Total	1,889	1,611	2,011

Source: KNTO 1995.

tourists. Among the businesses the average revenue of gift shops was US$1,082,000, followed by travel agencies (US$918,000), and restaurants (US$703,000). The data on average annual sales revenue for Korean tourist businessmen in Queensland were not available, however they appear similar to those of Hawaii, based on personal observation and discussion.

Tourist expenditures on shopping usually have a great impact on the economy in a tourist destination. When tourists purchase goods produced and made in the local area it accelerates the destination's economy by stimulating the output of related industries. Therefore, tourism authorities concerned with the destination usually show considerable interest in tourists' expenditures on shopping.

Korean tourists' favourite shopping items by category showed that in Hawaii jewellery was the most popular item followed by local products (macadamia nuts and chocolate), health foods (vitamins and honey) and clothes. In Queensland health foods (honey and vitamins) were the favourite item followed by local products (sheepskin and leather), jewellery (opal) and clothes. This indicates that there are some differences between the average Korean outbound tourists' tastes in different destinations (Table 15.6). This may be the result of a number of factors, including different knowledge of destinations, different markets, varying advertising and promotion and different availability of produce.

The average number of employees for Korean tourist businesses was 10.1 persons in Hawaii and 11.4 persons in Queensland, which indicates that the size of businesses in both destinations is much the same. The average number of employees in travel agencies was the highest of the categories with 13.4 persons in both destinations followed by gift shops (11.1 persons in both) and restaurants (5.1 persons in Hawaii and 9.2 persons in Queensland; Table 15.7).

The overall economic impact on residents in both destinations was shown to be positive. More that 70 per cent of total respondents answered that their revenue from Korean tourists had not only increased income (73.9 per cent

Table 15.6 Koreans' favourite shopping items by category

Category	*Hawaii Rank*	*Queensland Rank*	*All Koreans* Rank*
Jewellery	1	3	6
Local products	2	2	4
Healthy food	3	1	5
Clothes	4	4	3
Cosmetics	5	5	1
Whisky (spirits)	6	6	2

**Source*: KNTO 1995.

Table 15.7 Number of employees by category of business

Category	*Hawaii*	*Queensland*
Travel agency	13.4	13.4
Restaurant	5.1	9.2
Gift shop	11.1	11.1
Total	10.1	11.4

Source: Survey data.

for Hawaii and 72.4 per cent for Queensland), but also allowed the creation of new jobs (50.0 per cent for Hawaii and 52.7 per cent for Queensland), generated output of related industries (47.8 per cent for Hawaii and 48.8 per cent for Queensland), had a propaganda effect on local products (17.4 per cent for Hawaii and 16.2 per cent for Queensland), and had increased local taxes (12.0 per cent for Hawaii and 11.8 per cent for Queensland; Table 15.8).

Sociocultural impacts

In terms of sociocultural impacts, the respondents gave a negative reaction except with respect to cultural exchange effects (17.2 per cent for Hawaii and 16.4 per cent for Queensland). The main reasons for these negative reactions were the Korean tourists' discourteous behaviour (68.2 per cent for Hawaii and 52.4 per cent for Queensland), their lack of public morals (20.4 per cent for Hawaii and 62.0 per cent for Queensland), the cultural gap between visitors and residents (14.6 per cent for Hawaii and 24.2 per cent for Queensland), the tourists seeming too hurried (7.4 per cent for Hawaii and 6.6 per cent for Queensland), and tourists 'showing off' (4.2 per cent for Hawaii and 3.4 per cent for Queensland; Table 15.9).

A few respondents said that sometimes they felt ashamed to be Korean themselves because of the Korean tourists' impoliteness, mentioning such

Table 15.8 Economic impact on Korean residents

	Hawaii		*Queensland*	
Category	*Rank*	*%*	*Rank*	*%*
Increased income	1	(73.9)	1	(72.4)
Creation of new jobs	2	(50.0)	2	(52.6)
Generation of output of related industries	3	(47.8)	3	(48.8)
Propaganda effect of local products	4	(17.4)	4	(16.2)
Increased local tax	5	(12.0)	5	(11.8)
Improved infrastructure	6	(10.2)	6	(9.4)

Source: Survey data.

Table 15.9 Sociocultural impact on the Korean residents

	Hawaii		*Queensland*	
Category	*Rank*	*%*	*Rank*	*%*
Discourteous behaviour	1	(68.2)	2	(52.4)
Lack of public morals	2	(20.4)	1	(62.0)
Cultural gap	3	(14.6)	3	(24.2)
Seeming hurried and rushed	4	(7.4)	4	(6.6)
Showing off	5	(4.2)	5	(3.4)

Source: Survey data.

behaviour as their disorderly manners, lateness, propensity to bargain, being too noisy, drunkenness and smoking in non-smoking areas. The respondents thought their social status in the local community was degraded because of the Korean tourists' unreasonable attitudes compared with overseas tourists from other countries. Respondents thought non-Korean local residents considered a Korean tourist not as an individual but as 'an example of all Koreans'. Furthermore, a few respondents said that they were insulted by the wealthy Korean tourists who exaggerate and show themselves off in public.

The problem is that these phenomena can create not only negative relations between guests and hosts, but also can result in Korean tourists not being treated properly by Korean tourist businessmen in the destinations. Action needs to be taken, therefore, to correct this situation, as left untackled it is likely to become more severe, particularly in the face of rapidly increasing numbers of Korean tourists to these and other destinations.

Korean residents' attitude toward Korean tourists

When asked about their overall attitude toward Korean tourists, 26.2 per cent of the sampled Korean residents in Hawaii and 41.5 per cent of those in

Queensland said they welcomed Korean tourists. This indicates that Korean residents of Queensland are more receptive in their attitudes to the Korean tourists than those in Hawaii. Although residents do welcome Korean tourists, a sizable minority, 46.5 per cent of the sample in Hawaii and 39.1 per cent in Queensland, said Korean tourists are not welcome (Table 15.10).

The main reason they gave for welcoming Korean tourists was that the tourists bring money, create jobs and cause economic benefits. Other than the economic benefits, however, they gave no reason for them to welcome the Korean tourists. This suggests that the sociocultural impact of Korean tourists on the residents is negative. Furthermore, respondents in Hawaii said that only 19.6 per cent of the total Korean tourists showed interest in local cultural attractions and heritage, and more than 69.0 per cent of them did not care about the sociocultural resources in the destination (Kim 1993).

Although 5.3 per cent and 14.6 per cent respectively of Korean tourists to Hawaii and Queensland responded that tour guide service was poor, the residents blame the tourists for this attitude. The Korean residents think that improving the situation is completely dependent on changes in the attitude and behaviour or manner of Korean tourists visiting the destinations.

DISCUSSION

As previously stated, Korea faces two issues to be solved with respect to outbound tourism: first, to reduce the deficit in the international tourism balance of payments, and secondly, to improve the sociocultural impacts of Korean tourists on the residents at destinations.

Data collected show that an average Korean tourist businessmen in Hawaii earned US$875,000 annually from selling goods and services to Korean tourists, which ended up increasing household income and creating new jobs for many other local residents. As well, the additional expenditures of Korean tourists, such as to taxi drivers, supermarkets and food stores, all contribute to the local economy.

Table 15.10 The residents' attitude toward Korean tourists

Category	*Hawaii* %	*Queensland* %
Most welcome	4.3	4.5
Welcome	21.9	37.0
Neutral	27.1	19.6
Unwelcome	46.5	39.1

Source: Survey data.

Since the Korean tourists' spending power is greater than average tourist expenditure, it is obvious that the tourism income multiplier effect in the local community is greater than for the overseas tourists to Hawaii (Kim 1993). In 1995, the Hawaii State Government authority reported that 40.6 per cent of tourist expenditures contributed to the creation of household income (Department of Business, Economic Development and Tourism, State of Hawaii 1996). By applying the result of this study to the total amount of tourist expenditures of Koreans who visited Hawaii in 1995 (US$171,148,000), it can be seen that US$69,487,000 was contributed to the household income of Hawaiian residents (Table 15.11). In the case of Australia, 73.1 per cent of tourist expenditures contributed to the increase of household income (Commonwealth Department of Tourism, Australia 1994–5). By applying this figure to Korean tourist expenditures, it was discovered that a total amount of US$273,592,000 (US$168,985,000 in direct household income and US$104,607,000 in indirect household income) was contributed to increase household income (Table 15.11).

In Hawaii in 1995, it was reported that employment opportunities were provided to the extent of 1.4 persons per US$100,000 of tourist expenditures (Department of Business, Economic Development and Tourism, State of Hawaii 1996). This indicates that Korean tourist expenditures created a total of 2,387 jobs in 1995 (Table 15.12). In Australia the analysis showed that employment opportunities were provided at 2.7 persons per US$100,000 of tourist expenditures (Commonwealth Department of Tourism, Australia 1994–5), which indicates that Korean tourist expenditure created an effective increase of employment opportunities for 10,081 people in Australia.

Consequently, the impact of Korean tourism on Korean tourist businessmen in the destinations is not only positive in direct economic terms, but also provides the benefit of household income and effective employment opportunities to many of the residents. This effectiveness is a very important

Table 15.11 Impact of tourist expenditure for 1995 on household income

			Household income		
Category	*No. of tourists*	*Tourist expenditure US$000*	*Direct US$000*	*Indirect US$000*	*Total US$000*
Hawaii*	104,550	171,148	–	–	69,487
Australia**	174,000	388,107	175,047	108,659	283,706

Source: *Department of Business, Economic Development and Tourism, State of Hawaii, 1st Quarter Hawaii's Economy, 1996, pp. 6–7.
** Commonwealth Department of Tourism, Australia, Annual Report, 1994–5, pp. 9–10.

Table 15.12 Impact of tourist expenditure for 1995 on employment

		Employment			
Category	*Tourist expenditure*	*Direct US$000*	*Indirect US$000*	*Total US$000*	*Employment/ $100,000 tourist expenditure*
Hawaii**	171,148	–	–	2,387	1.4
Australia*	388,107	7,305	3,138	10,443	2.7

Sources: *Department of Business, Economic Development and Tourism, State of Hawaii, 1st Quarter Hawaii's Economy, 1996, pp.6–7.
** Commonwealth Department of Tourism, Australia, Annual Report, 1994–5, pp. 9–10.

factor in economic development in the destinations. As the above analysis shows, the economic effects of tourism are positive in both destinations, Hawaii and Australia. In comparison, it was shown that Australia has gained greater economic benefits than Hawaii.

In order to increase the economic benefits of tourism, both Korean tourist businessmen and the tourist authorities in the two countries have been endeavouring to attract Korean tourists. As result of such efforts, it has been shown that during the last seven years, Australia attracted 560 per cent more Korean tourists than Hawaii, despite being a considerably greater distance away. In addition, attitudes of the sample in Queensland are less negative towards Korean tourists than in Hawaii, perhaps because Korean tourism to Australia is more recent than to Hawaii.

In order to continue to attract Korean tourists and to maximize the economic benefits from them, Korean businessmen in the destinations must also address problems experienced by the tourists. These problems have been termed 'the inconvenience factor' and surveyed by KNTO and the author (Kim 1993, 1996).

The inconvenience factors experienced by Korean overseas tourists were arbitrarily segmented by KNTO for the survey on Korean overseas travel (Table 15.13). According to Kim's studies (1993, 1996), in the case of Hawaii and Queensland, the most inconvenient factor was communication (84.2 per cent for Hawaii and 85.4 per cent for Queensland), followed by local food problems (51.9 per cent for Hawaii and 38.6 per cent for Queensland), transportation (29.8 per cent for Hawaii and 53.2 per cent for Queensland), and travel information (16.4 per cent for Hawaii and 40.8 per cent for Queensland). These top four inconvenience factors remain the same for all Korean overseas tourists to all international tourist destinations. In order to cope with these adverse factors, it is suggested that tourism businessmen in both destinations should make an effort to enable Korean tourists to overcome the cultural differences between themselves and local

Table 15.13 Inconvenience factors ranked by the Korean tourists

Category	*Hawaii** *Rank*	*%*	*Queensland*** *Rank*	*%*	*Overall (Including all destinations)* *Rank*	*%*
Language barrier	1	(84.2)	1	(85.4)	1	(61.7)
Food service	2	(51.9)	4	(38.6)	2	(44.8)
Transport system	3	(29.8)	2	(53.2)	3	(24.8)
Travel information	4	(16.4)	3	(40.8)	4	(18.9)
Currency exchange	5	(13.4)	5	(16.2)	7	(8.7)
Flight reservation	6	(10.1)	7	(10.2)	5	(11.6)
Night life/entertainment	7	(6.5)	8	(8.4)	11	(3.1)
Tour guide service	8	(5.3)	6	(14.6)	8	(7.6)
Forced tipping	9	(4.8)	10	(6.9)	9	(6.9)
Forced shopping	10	(4.0)	9	(7.4)	10	(5.7)
Security service	11	(3.4)	11	(5.2)	6	(9.4)

Source: Kim, S.M. (*1993 and **1996).

residents and to assist local tourism authorities to understand Korean culture at the same time.

On the other hand, as shown by the analysis of the empirical study, impolite behaviour and cultural differences between Korean tourists and local residents have resulted in negative responses towards the tourists, which is a significant problem. It becomes an especially significant problem when local residents say that they regard a Korean tourist not as 'an individual' but as 'a Korean'. It has not been fully appreciated that the indiscreet behaviour of some Korean tourists can be seen as a common standard by which all Korean people are judged, and it has resulted in a deterioration of the Korean national image.

It is necessary, therefore, to solve these problems. Tourists and local residents need to accept and broaden the scope of understanding of each other. Korean tourist businessmen, because of their unique position as intermediaries in this situation, as noted earlier, should endeavour to minimize the negative effects of tourism by helping Korean tourists understand local reactions, and by persuading them to recognize and minimize the cultural differences between themselves and local residents.

CONCLUSIONS

The questions raised in this study are: first, the Korean government's concerns about a deficit in international balance of tourism payments; second, the negative sociocultural impact of Korean tourists on Korean residents in destinations; third, the inconvenience factors experienced by Korean tourists; and fourth, the distorted perspective of local residents in tourist destinations about the Korean people as a whole.

Accordingly, the possible solutions and recommendations for improving these issues can be suggested as follows:

1 With respect to the economic aspect, the average Korean tourist businessman in Hawaii repatriates about US$10,000 back to Korea annually; gift shop owners send US$12,000 back to Korea, followed by travel agency owners (US$11,000) and restaurant owners (US$9,000). This information is not exact since the Korean businessmen were reluctant to provide detailed financial information. For the same reason, the monetary amounts repatriated to Korea were not obtained in the case of the Queensland sample. Consequently, the author was unable to identify the leakage of their profits. Various products imported from Korea are sold at gift shops and Korean food stores in the tourist destinations. These items should be considered in the international balance of tourism payments. Therefore, it is not effective to analyse only what is shown in the visible international balance of tourist expenditures in relation to the balance of payments. By taking into account the effects of return currency, and exports for tourist purchases overseas, a new device for calculating the international balance of tourism payments should be sought. A new calculation of international tourism balance of payments is needed which takes these factors into account in order to provide a more accurate estimate of the true international balance of tourism payments.
2 In order to minimize the negative effects of the sociocultural stereotype which influences local residents' attitudes towards Korean tourists, it is necessary that the tourists themselves make an effort to obtain cultural information on destinations prior to departure from Korea. Simultaneously, Korean tourist businessmen in destinations need to enlighten and persuade Korean tourists to reduce their improper behaviour. Ideally, tourists and locals should have an opportunity to meet and to demonstrate their own values and customs, and for tourists to enjoy the cultural attractions of the destination through contact with local residents. Through appropriate arrangements, both tourists and residents could maximize the opportunity for the natural sociocultural exchange which international tourism allows. Korean tourist businessmen and tourism suppliers should also improve their own attitudes before they impute the sole responsibility for negative sociocultural influences to the Korean tourists.
3 A positive attitude toward reducing Korean tourists' inconveniences, such as problems in communication, and with local food, transportation, travel information and money exchange is required among the tourist

businessmen in destinations. The ability to remove or reduce these impediments to international tourism is dependent not only upon the tourists' own endeavours to adapt to strange circumstances but also a matter of concern for the local tourist authorities: Korean tourist businessmen should therefore co-operate with these bodies in order to devise and implement appropriate solutions and improvements.

4 It is not acceptable that local residents should treat Koreans as a homogeneous group. In other words, the Korean tourists' behaviour should not be considered as representative of 'all Koreans' consciousness'. They must be treated as 'individuals' so that misinterpretation and misunderstanding caused by cultural differences can be reduced or eliminated. Therefore, the biased perception of the local residents should be changed along with the Korean tourists' attitudes and behaviour toward local residents and tourist business proprietors. It is necessary that the best efforts should be made by the both tourists and residents as international citizens to promote a good image to each other.

In the immediate future at least Korean participation in international tourism can be expected to decrease sharply due to the depreciation of the Korean currency because of the economic recession which began in 1997. As a result of this, Korean tourists are likely to have less economic impact on particular tourist destinations while more positive sociocultural impacts can be expected in the immediate future. Research on the results of economic recession on destination areas through integrated impact analysis is a topic which should be of interest in the foreseeable future, not only in the case of Korean tourism, but for all of the South-East Asian countries affected by the current economic downturn in their economies.

References

Ap, J. (1990) 'Residents' perceptions research on the social impacts of tourism', *Annals of Tourism Research* 17 (4): 610–15.

Archer, B.H. (1973) *The Impact of Domestic Tourism*, Cardiff: University of Wales Press.

—— (1977) *Tourism Multipliers: The State of the Art*, Cardiff: University of Wales Press.

—— (1982) *The Economic Impact of Tourism in Seychelles*, Commonwealth Secretariat, CFTC/SEY/51.

Archer, B.H. and Wanhill, S.R.C. (1981) *The Economic Impact of Tourism in Mauritius*, World Bank, MAR/80/004/A/01/42.

—— (1985) *Tourism in Bermuda: An Economic Impact Study*, Government of Bermuda.

Belisle, F.J. and Hoy, D. (1980) 'The perceived impact of tourism by residents; a case study in Santa Marta, Colombia', *Annals of Tourism Research* 7 (1): 83–101.

Commonwealth Department of Tourism (1994–1995) Australia Annual Report, 9–10.

Department of Business, Economic Development and Tourism, State of Hawaii (1996) 1st Quarter Hawaii's Economy, 6–7.
Fletcher, J.E. (1987) *The Economic Impact of International Tourism on the National Economy of the Solomon Islands*, WTO/UNDP.
Hawaii Visitors Bureau (1990–1995) *Annual Research Report*, Honolulu.
Johnson, J.D., Snepenger, D.J. and Akis, S. (1994) 'Residents' perceptions of tourism development', *Annals of Tourism Research* 21 (3): 629–42.
Kim, S.M. (1993) 'A study on the actual conditions of Korean outbound tourism and recommendations for its improvements', *Study on Tourism*, 17, Korea.
—— (1996) 'A study on the actual conditions of Korean tourists to Australia and recommendations for improvements', *Tourism Study*, 6, Korea.
Korea National Tourism Organization (1990–1995) *Annual Tourism Report*, Seoul, Korea.
—— (1990–1996) *Tourism Statistical Report*, Seoul, Korea.
Mathieson, A. and Wall, G. (1986) *Tourism: Economic, Physical and Social Impacts*, London: Longman.
Mill, R.C. and Morrison, A.M. (1985) *The Tourism System: An Introductory Text*, Prentice–Hall International Editions, p. 221.
Reynolds, P.C. (1992) 'Impacts of tourism on indigenous communities: the Australian case', pp. 113–19 in C.P. Cooper and A. Lockwood (eds) *Progress in Tourism, Recreation and Hospitality Management 4*.
Sadler, P., Archer, B.H. and Owens, C. (1973) *Regional Income Multipliers*, Cardiff: University of Wales Press.
Var, T., Kendall, K.W. and Tarakcioglu, E. (1985) 'Resident attitudes toward tourists in a Turkish resort town', *Annals of Tourism Research* 12 (4): 652–8.
Wanhill, S.R.C. (1983) 'Measuring the economic impact of tourism', *Service Industries Journal* 3 (1): 9–20.
—— (1988) Tourism multipliers under capacity constraints, *Service Industries Journal* 8 (2): 136–42.

INDEX